SpringerBriefs in Electrical and Computer Engineering

More information about this series at http://www.springer.com/series/10059

Tianyu Wang · Lingyang Song
Walid Saad · Zhu Han

Overlapping Coalition Formation Games in Wireless Communication Networks

 Springer

Tianyu Wang
School of Electrical Engineering
 and Computer Science
Peking University
Beijing
China

Lingyang Song
School of Electrical Engineering
 and Computer Science
Peking University
Beijing
China

Walid Saad
Wireless@VT, Bradley Department
 of Electrical and Computer Engineering
Virginia Tech
Blacksburg, VA
USA

Zhu Han
Electrical and Computer Engineering
 Department
University of Houston
Houston, TX
USA

ISSN 2191-8112   ISSN 2191-8120   (electronic)
SpringerBriefs in Electrical and Computer Engineering
ISBN 978-3-319-25698-6   ISBN 978-3-319-25700-6   (eBook)
DOI 10.1007/978-3-319-25700-6

Library of Congress Control Number: 2016949113

Printed on acid-free paper

This Springer imprint is published by Springer Nature
The registered company is Springer International Publishing AG
The registered company address is: Gewerbestrasse 11, 6330 Cham, Switzerland

Preface

Cellular networks are witnessing an unprecedented evolution from classical, centralized, and homogenous architectures into an extreme complex network structure, in which a large number of network devices are densely deployed in an irregular, decentralized, and heterogenous fashion. This shift in network architecture requires network devices to become more flexible, autonomous and cooperative, so as to meet the various performance requirements of different wireless services.

In this book, we focus on communication systems in which network devices need to coordinate with each other so as to increase the overall performance. Such coordination can, for example, take place between small access points that seek to coordinate their radio resource allocation in the same spectrum, ultradense small cells with massive MIMO that share the same limited pilot sequences, mobile users with nonorthogonal multiple access (NOMA) that share the same subchannels, nearby single-antenna users that can cooperatively perform virtual MIMO communications, or even unlicensed users that wish to cooperatively sense the spectrum of the licensed users.

Conventionally, the solutions of these cooperative scenarios have been based on either centralized optimization algorithms that require a large amount of computational resources (e.g., subchannel allocation in NOMA), or decentralized heuristic methods with suboptimal performances (e.g., pilot reuse in Massive MIMO). More recently, there has been a surge in models that adopt the framework of *cooperative games*, in which the coordination among network devices is formulated by the coalition formation process performed by self-interested players. These game theoretical methods typically seek to establish distributed solutions that are stable, in the sense that no device has an incentive to change its coalition membership. Their outcome can achieve a balance between the computational complexity and the network performance. Indeed, cooperative games, in general, and *coalition formation games* (CF games), in particular, have become a popular tool for analyzing wireless networks.

Most of the existing body of work focuses on coalition formation models in which the players form separate coalitions and achieve performance gain from the single coalition they join. However, in many cooperative scenarios of future wireless systems, a network device may need to join multiple groups of cooperative devices, and thus, these cooperative groups may overlap with each other. For example, in the subchannel allocation problem in NOMA, the mobile users share the same subchannel form a coalition, and thus, each mobile user join multiple coalitions as they access multiple subchannels. In this book, we introduce a mathematical framework from cooperative games, known as *overlapping coalition formation games* (OCF games), which provides the necessary analytical tools for analyzing how network devices in a wireless network can cooperate by joining, simultaneously, multiple overlapping coalitions. Therefore, the readers can utilize this book as a tool to address the cooperative scenarios in future wireless networks.

First, in Chap. 1, we introduce the basic concepts of CF games and OCF games in general, and develop two polynomial algorithms for two classes of OCF games, i.e., K-coalition OCF games and K-task OCF games, respectively. Then, in Chaps. 2 and 3, we present two emerging applications of OCF games in small cell-based heterogeneous networks (HetNets) and cognitive radio networks, in order to show the advantages of forming overlapping coalitions compared with the traditional nonoverlapping CF games. Finally, in Chap. 4, we discuss the potential challenges of using OCF games in future wireless networks and briefly present some other potential applications.

Acknowledgement

This work was supported by US NSF ECCS-1547201, CCF-1456921, CNS-1443917, ECCS-1405121, and NSFC61428101, and by the U.S. National Science Foundation under Grant AST-1506297.

Beijing, China   Tianyu Wang
Beijing, China   Lingyang Song
Blacksburg, USA   Walid Saad
Houston, USA   Zhu Han

Contents

Abbreviations

BS	Base station
CF games	Coalition formation games
CoMP	Coordinated multipoint
CR	Cognitive radio
CSS	Cooperative spectrum sensing
DCS	Distributed cooperative sensing
DSL	Digital subscription line
FC	Fusion center
FDD	Frequency division duplexing
HetNets	Heterogeneous networks
MBSs	Macro base stations
MIMO	Multiple input multiple output
MUEs	Macrocell user equipments
NOMA	Nonorthogonal multiple access
NTU	Nontransferable utility
OCF games	Overlapping coalition formation games
OFDMA	Orthogonal frequency division multiple access
PUs	Primary users
RATs	Radio access technologies
SBSs	Small cell base stations
SU	Secondary users
SUEs	Small cell user equipments
TU	Transferable utility

Chapter 1
Introduction

In this chapter, we formally introduce the notions of coalition formation games (CF games) and the extended overlapping coalition formation games (OCF games). In particular, we present the basic game models and stability notions. For OCF games, we show that computation of stable outcomes can generally be intractable, and thus, we identify several constraints that lead to tractable subclasses of OCF games, and provide efficient algorithms for solving games that fall under these subclasses.

1.1 Coalition Formation Games

1.1.1 Game Models

Game theory is a mathematical framework that can be used to analyze systems that involve multiple decision makers having interdependent objectives and actions. The decision makers, which are usually referred to as players, will interact and obtain individual profits from the resulting outcome. In cooperative games, the players can form cooperative groups, or coalitions, to jointly increase their profits in a game. In this section, we present an important type of cooperative games, *coalition formation games* (CF games), in which the players are typically assumed to form multiple, disjoint, and nonoverlapping coalitions, and a player only cooperates with players within the same coalition. In CF games, if the cooperative value generated by a coalition can be quantified by a real number, and this gain can be divided in any manner among the coalition members, these games are referred to games with transferable utility (TU) [1]. If there exists some rigid restrictions on the distribution of coalition value, e.g., the payoff of a coalition member is determined by the action profile of all the coalition members, these games are referred to as games with nontransferable utility (NTU) [2, 3]. We formally give the definitions of games with TU and NTU as follows.

© The Author(s) 2017
T. Wang et al., *Overlapping Coalition Formation Games in Wireless Communication Networks*, SpringerBriefs in Electrical and Computer Engineering, DOI 10.1007/978-3-319-25700-6_1

Definition 1 A coalition formation game with a transferable utility is defined by a pair $(\mathcal{N}, v)$, where $\mathcal{N}$ is the set of all N players and $v : 2^N \rightarrow \mathbb{R}$ is a value function, such that for every coalition $S \subseteq \mathcal{N}$, $v(S)$ is the amount of value that coalition S receives from their cooperation, which can be distributed in any arbitrary manner among the members of S. We denote by x_i as the payoff of player $i \in S$, and thus, we have $\sum_{i \in S} x_i = v(S)$.

Definition 2 A coalition formation game with a nontransferable utility is defined by a tuple $(\mathcal{N}, \{x_i\}_i)$, where $\mathcal{N}$ is the set of all N players and $x_i : 2^N \rightarrow \mathbb{R}$ is a payoff function, such that for every coalition $S \subseteq \mathcal{N}$, $x_i(S)$ is the amount of payoff that player $i \in S$ receives from its coalition S.

We note that there are some other game models that are related to CF games. If the value function satisfies the superadditive property, the players always form a grand coalition that involves all N players, and the games are referred to as canonical coalitional games. If the value function is determined by a graph structure within the coalition, the games are referred to as coalitional graph games. For readers interested these concepts, we refer to the survey [4] and the book [5].

1.1.2 Core of Coalition Formation Games

In CF games, the outcome is represented by a partition of the set of players $\mathcal{N}$, which is referred to as coalition structure CS, and the payoff allocation $\mathbf{x} = (x_1, \ldots, x_N)$ of all the players. There exist several notions of stability for CF games, which includes the *core* [6], the *Shapley value* [2], and *nucleolus* [7]. In this book, we adopt the most well-studied core notion to study the stability of an CF game. Specifically, the core is a set of payoff allocations that guarantees that no group of players has an incentive to leave the current coalition structure CS and form another coalition $S \notin CS$. For CF games with TU, it is defined as

$$\mathcal{C}_{TU} = \left\{ \mathbf{x} : \sum_{i \in S} x_i = v(S), \forall S \in CS \text{ and } \sum_{i \in S} x_i \geq v(S), \forall S \notin CS \right\} \quad (1.1)$$

For CF games with NTU, it is defined as

$$\mathcal{C}_{NTU} = \{ \mathbf{x} : x_i = x_i(S), \forall i \in S, S \in CS \text{ and}$$
$$\exists i \in S, x_i > x_i(S), \forall S \notin CS \} \quad (1.2)$$

In other words, if an outcome lies in the core, then for any set of players $S \subseteq \mathcal{N}$ that intend to reject the proposed payoff allocation $\mathbf{x}$, deviate from the current coalition structure CS and form a new coalition $S \notin CS$, there must be at least one player i whose new payoff x_i' is decreased, i.e., $x_i' < x_i$. In general, finding an coalition structure CS and an associated stable payoff allocation $\mathbf{x}$ in the core, is NP-hard, as the number of partitions of a set $\mathcal{N}$ grows exponentially with the number of players N, which is given by a value known as the Bell number [8]. However, there exist several distributed approaches that aim to solve the coalition formation problem by using weaker stability notions [8–12].

The most popular approach is the merge-and-split algorithm [10, 12], in which the players form coalition structures by following two simple rules, the merge rule and the split rule. Specifically, given a collection of disjoint coalitions $\{S_1, S_2, \ldots, S_L\}$, the players in $S = \uplus_{l=1}^{L} S_l$ agree to merge into a single coalition S, if the new coalition is preferred by all the players in S, i.e., $\sum_{l=1}^{L} v(S_l) \leq v(S)$ for TU and $x_i(S_l) \leq x_i(S), \forall i \in S_l$ for NTU. Similarly, a coalition S splits into multiple disjoint coalitions $\{S_1, S_2, \ldots, S_L\}$, if the resulting collection is preferred by all the players in S. It has been proved that the merge-and-split algorithm can converge [10, 12], and the resulting outcome satisfies a weak equilibrium-like stability, known as $\mathbb{D}_{hp}$ stability, which simply implies that no group of players has an interest in performing a merge or a split operation. Clearly, if the outcome is in the core, it must also satisfy $\mathbb{D}_{hp}$ stability, but the inverse is not true, i.e., $core \subset \mathbb{D}_{hp}$.

1.2 Overlapping Coalition Formation Games

1.2.1 Game Models

In many practical cooperative scenarios, the players may be involved in multiple coalitions simultaneously. In such cases, these players may need to split their resources among the coalitions in which they are participating. Consequently, some of the coalitions may involve some of the same players thereby overlapping with one another. Next, we formally introduce the mathematical tool to model these "overlapping" situations, *cooperative games with overlapping coalitions*, or *OCF games* [13].

In OCF games, each player possesses a certain amount of resources, such as time, power, or money. In order to obtain individual profits, the players form coalitions by contributing a portion of their resources and receive payoffs from the devoted coalitions. A *coalition* can be represented by the resource vector contributed by its coalition members, i.e., $r = (r_1, r_2, \ldots, r_N)$, where $0 < r_i < R$ represents player i's resources that are contributed to this coalition. For each coalition r, the *coalition value* is decided by a function $v : [0, R]^N \to \mathbb{R}^+$, which represents the total payoff that the players can get from a cooperative coalition. The coalition value can be divided to the coalition members based on specific rules, e.g., the value can be equally divided among coalition members, or it can be divided based on the contribution of each member. We denote by x as the payoff allocation rule, and accordingly, the individual payoff that player i receives from coalition r is denoted by $x_i(r)$. The players may decide to devote different amount of resources into different coalitions, so as to maximize its individual payoff $p_i(\pi, x) = \sum_{r \in \pi} x_i(r)$, where π represents the set of all coalitions $\pi = \{r^1, r^2, \ldots, r^K\}$ formed by the players. Note that these

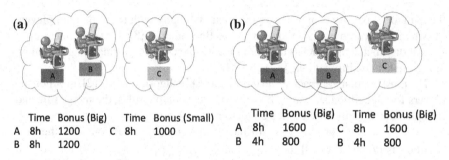

(a)

	Time	Bonus (Big)		Time	Bonus (Small)
A	8h	1200	C	8h	1000
B	8h	1200			

(b)

	Time	Bonus (Big)		Time	Bonus (Big)
A	8h	1600	C	8h	1600
B	4h	800	B	4h	800

Fig. 1.1 An example to show the differences between OCF games and traditional CF games

coalitions may have common members, and thus, they form an *overlapping coalition structure*. Compared with the traditional CF games, OCF games allow the players to form an overlapping coalition structure and get payoffs from multiple coalitions. Therefore, the overlapping structure may provide more flexibility for the players to utilize their resources, which enables the coalitions to be better organized and potentially leads to outcomes with higher payoffs. The following example clearly shows the potential advantage of OCF games.

Example: Consider a software company with three developers A, B, and C. Each developer works 8 h a day. There are two types of projects in the company, big projects and small projects. A big project requires 12 man-hours per day and provides a 2400 bonus, and a small project requires 8 man-hours per day and provides a 1000 bonus. We assume the bonus is divided to the participating developers according to their devoted time. In a traditional CF game, the players can only form disjoint coalitions, and the optimal coalition structure is $\{\{A, B\}, \{C\}\}$, as seen in Fig. 1.1a, i.e., developers A and B work together to accomplish a big project and developer C work alone for a small project. The total payoffs of A, B, and C are then given by (1200, 1200, 1000). In an OCF game, the players can split their time into different coalitions, and the optimal coalition structure is $\{(8, 4, 0), (0, 4, 8)\}$, as seen in Fig. 1.1b, i.e., developers A and B devote 8 and 4 h to accomplish a big project, and developers B and C devote 4 and 8 h to accomplish another big project. The payoffs of A, B, and C are then given by (1600, 1600, 1600).

1.2.2 *A-core of Overlapping Coalition Formation Games*

In cooperative games, one must seek a stable outcome, i.e., a coalition structure in which no set of players can deviate and obtain a new structure that increases all their payoffs. In traditional CF games, the deviating players, or deviators, leave their original coalitions and form a new coalition, the value of which is the total payoff that the deviators can get from their deviation. If there exists a payoff division that makes all the deviators achieve a higher payoff compared with the original coalition structure, we say the deviation is profitable. If there exists no profitable deviation, we say the structure is stable. Particularly, if the players have an incentive to form the grand coalition that includes all players, the cooperative game is called a canonical game and the set of all stable payoff divisions, corresponding the grand coalition, is called the core of such a game.

Compared with CF games, defining notions for stability in OCF games is more complicated due to the overlapping property. In OCF games, instead of forming a single coalition, the deviators may form multiple coalitions that overlap one another thus complicating the computation of the maximal total payoff of such an overlapping coalition structure. Also, the deviators in OCF games may partially deviate from the original coalitions by withdrawing a portion of their resources while maintaining their other resources in their original coalitions. Therefore, one must precisely define how the nondeviators in the original coalitions will react to such a deviation and how much payoff the deviators can get from those partially deviated coalitions. To this end, we now present the $\mathcal{A}$-core, a stability notion suitable for OCF games, which is an extension of the core notion from traditional CF games.

We begin by defining a deviation in OCF games. Given a coalition structure π and a set of players S that attempts to deviate the structure, the coalitions in π can be divided into two groups: the coalitions that only involve players in S, denoted by $\pi|_S$, and the coalitions that involves players other than S, given by $\pi \setminus \pi|_S$. Since coalitions $\pi|_S$ are fully controlled by the deviators, they should be seen as pure resources withdrawn by the deviators. While coalitions $\pi \setminus \pi|_S$, which include both deviators and nondeviators, should be considered as coalitions that are partially deviated by the deviators. We define $D(\pi) = \{d(r) | r \in \pi \setminus \pi|_S\}$ as the resources withdrawn from $\pi \setminus \pi|_S$, where $d(r)$ is the resources for coalition r.

The deviators S can form an overlapping coalition structure using both the withdrawn resources $D(\pi)$, and the resources of their own $\pi|_S$. We denote by W_S as the sum available resources of the deviators S, and $\Pi(W_S)$ as the set of all possible coalition structures that can be formed using W_S. The optimal coalition structure formed by S is then given by $\pi(W_S) = argmax_{\pi \in \Pi(W_S)} \{\sum_{r \in \pi} v(r)\}$. Note that the deviators may also receive payoffs from the coalitions that they partially deviate, i.e., $\pi \setminus \pi|_S$. We formally define the *arbitration function* $\mathcal{A}_r(\pi, x, D(\pi), S)$, which represents the total payoff that the deviators S will receive from coalition $r \in \pi \setminus \pi|_S$.

Definition 3 A deviation on a player set S is said to be $\mathcal{A}$-*profitable* if and only if

$$\sum_{i \in S} p_i(\pi, x) < \sum_{r \in \pi(W_S)} v(r) + \sum_{r \in \pi \setminus \pi|_S} \mathcal{A}_r(\pi, x, D(\pi), S). \qquad (1.3)$$

If there exists no $\mathcal{A}$-profitable deviation for any player set, then the coalition structure π is said to be in the $\mathcal{A}$-*core*, or $\mathcal{A}$-*stable*.

The coalition structures in $\mathcal{A}$-core represent the stable structures in which no set of players have the motivation to deviate from the current structure. We note that the definition of $\mathcal{A}$-core depends on the specific form of the arbitration function.

According to different assumptions on players and different applicable scenarios [13–15], three of the mostly used arbitration functions are described as follows:

1. *c-core*: if the players are very conservative in cooperating with deviators, the deviators may receive no payoffs from any original coalitions in $\pi \setminus \pi|_S$, even if they still contribute to these coalitions, i.e., $\mathcal{A}_r(\pi, x, D(\pi), S) \equiv 0$. This is called the *conservative arbitration function*, and the stability notion is referred to as *c-core*.

2. *r-core*: if the players are more lenient, the deviators can still get their original payoffs from the coalitions that are not influenced by their deviation, i.e., $\mathcal{A}_r(\pi, x, D(\pi), S) = \sum_{i \in S} x_i(r)$ for all coalition r with $d(r) = 0$. This is called the *refined arbitration function*, and the stability notion is referred to as *r-core*.

3. *o-core*: the players can be highly generous that they allow the deviators to keep all the "leftover" payoff as long as the nondeviators' original payoffs are ensured to be unchanged, i.e., $\mathcal{A}_r(\pi, x, D(\pi), S) = v(r - d(r)) - \sum_{i \in \mathcal{N} \setminus S} x_i(r)$. This is called the *optimistic arbitration function*, and the stability notion is referred to as *o-core*.

Hereinafter, we adopt the *o-core* as the stability notion due to its computational advantage that we will explain in the following section.

1.3 Computing o-Stable Outcomes for OCF games

1.3.1 *o-Stable Outcomes*

To avoid the difficulty in representing the value function v and the arbitration function $\mathcal{A}$, we assume that the resources are given by integers such that $R \in \mathbb{Z}^+$ and the players can only divide their resources in a discrete manner. Such games are referred to as *discrete OCF games* and they apply to practical systems. Also, due to the cost of information exchange between deviators, we assume the number of deviators in a deviation is bounded. We denote by S the upper bound of deviation size, i.e., $|S| \leq S$ for any deviation on any player set S. It has been shown that computing an $\mathcal{A}$-stable outcome of an OCF game is generally a challenging problem [15]. However, if we only consider the o-core notion and identify several constraints on the game, there exist efficient algorithms that lead to o-stable outcomes of such games.

Proposition 1 *Any o-profitable deviation on any player set $S \subseteq \mathcal{N}$ will increase the social welfare, which is defined as the total value of all coalitions in the outcome, i.e., $\sum_{r \in \pi} v(r)$.*

The proof of the above proposition can be found in [16]. In most practical problems, the social welfare is bounded due to the limited resources. For such games, Proposition 1 implies that the game must converge to an o-stable outcome after finite o-profitable deviations. Therefore, we can compute an o-stable outcome by itera-tively computing an o-profitable deviation to the current outcome. However, finding an o-profitable deviation is a challenging problem. First, without further restrictions on deviations, the number of potential deviations can be extremely large. Second, deciding whether a deviation is o-profitable requires solving the optimal coalition structure problem, which is not a straightforward problem. Now, we define two sub-classes of OCF games, namely *K-coalition OCF games* and *K-task OCF games*, and we provide efficient algorithms to compute o-stable outcomes in such games, respectively.

1.3.2 K-coalition OCF Games

In a K-coalition OCF game, we assume each player can contribute to at most K coalitions. This assumption is reasonable in many practical systems due to geo-graphical constraints, communication cost, or lack of information. For example, a mobile user can only connect to limited base stations that are close to it. Therefore, for a group of deviators S in a K-coalition game, the number of possible devia-tions is bounded by $(R+1)^{SK}$. Since there are C_N^S groups of possible deviators, the total number of possible deviations is then bounded by $C_N^S(R+1)^{SK} = \mathcal{O}(N^S)$, which is polynomial in N. For any deviation on S, the deviators need to calculate the optimal coalition structure $\pi(W_S)$ to decide whether the deviation is o-profitable. We define the *superadditive cover* of v to be the function $v^* : [0, R]^N \to \mathbb{R}^+$, such that $v^*(W) = \max_{\pi \in \Pi(W)} \{\sum_{r \in \pi} v(r)\}$ for any resource vector W. Briefly, $v^*(W)$ is the maximal total value that the players can generate by forming over-lapping coalitions when their total resources are given by W. We observe that $v^*(W) = \max \{v^*(W-r) + v(r) | r \preccurlyeq W\}$, which is a recurrence relation for a discrete-time dynamic system. Thus, we can use the dynamic programming algorithm to calculate $v^*(W)$. Given the values of $v^*(W')$ for all $W' \preccurlyeq W$, the computation of $v^*(W)$ requires $(R+1)^S$ times of computing v. Therefore, the entire computation of $v^*(W)$ requires at most $(R+1)^S(R+1)^S = (R+1)^{2S}$ times of computing v. When $v^*(W)$ is calculated, we can trace backward the optimal path and achieve every coalition in the optimal coalition structure $\pi(W)$. Therefore, the optimal coalition structure $\pi(W_S)$ can be calculated in time $(R+1)^{2S}$. Therefore, we can calculate an o-profitable deviation in time $C_N^S(R+1)^{SK}(R+1)^{2S} = C_N^S(R+1)^{S(K+2)} = \mathcal{O}(N^S)$, which is polynomial in N. The algorithm for K-coalition games is shown in Table 1.1.

1.3.3 K-task OCF Games

In a K-task OCF game, each coalition in the game corresponds to a specific task and each player can only contribute to K tasks. Being different from K-coalition OCF games, the number of coalitions in a K-task OCF game is strictly limited by the

Table 1.1 Algorithm for K-coalition games

Input an initial outcome (π_0, x_0).

1: $\pi \leftarrow \pi_0$ % initial coalition structure
2: $x \leftarrow x_0$ % initial payoffs
3: **while** there exists an o-profitable deviation on a player set S **do**
4:    $(r - d(r)) \leftarrow r$ for all $r \in \pi \backslash \pi|_S$
5:    $\pi(W_S) \leftarrow \pi|_S$
6:    decide new payoffs x
7: **end while**
8: $\pi_f \leftarrow \pi$ % final coalitional structure
9: $x_f \leftarrow x$ % final payoffs

Output an o-stable outcome (π_f, x_f).

Table 1.2 Algorithm for K-task games

Input an initial outcome (π_0, x_0).

1: $\pi \leftarrow \pi_0$ % initial coalition structure
2: $x \leftarrow x_0$ % initial payoffs
3: **while** there exists an o-profitable transfer on a player set S **do**
4:    $(r - d(r)) \leftarrow r$ for all $r \in \pi$
5:    decide new payoffs x
6: **end while**
7: $\pi_f \leftarrow \pi$ % final coalitional structure
8: $x_f \leftarrow x$ % final payoffs

Output an o-stable outcome (π_f, x_f).

number of tasks, which are predetermined by the considered problem. For example, in a software company, the available projects are predetermined and the developers cannot form coalitions to generate new projects but only divide his time among the existing ones. Since the number of coalitions is fixed in a K-task OCF game, a deviation will not form new coalition structures but only move resources among the existing coalitions, and thus, we refer to deviation as *transfer* in K-task OCF games. The number of possible transfers is now given by $C_N^S [K^2(R+1)]^S = \mathcal{O}(N^S)$, which is polynomial in N. Since the deviators do not form an overlapping coalition structure, their payoffs can be easily calculated using the arbitration function. Therefore, an o-profitable deviation of a K-task OCF game can be calculated in time $\mathcal{O}(N^S)$. The algorithm for K-task games is shown in Table 1.2.

Given the polynomial algorithms that can be used to solve K-coalition games and K-task games, we then provide some applications to show how the concepts and algorithms of OCF games can be utilized in wireless networks. Note that we restrict our model to single-resource scenarios in which the players only have one type of resources. However, this model can be extended to the multiresource setting, by using a vector rather than a scalar to describe the contribution of a player, and all the concepts and algorithms can also be extended to such a case.

References

1. J. von Neumann, O. Morgenstern, *Theory of Games and Economic Behavior* (Princeton University Press, Princeton, 1944)
2. R.B. Myerson, *Game Theory, Analysis of Conflict* (Harvard Univeristy Press, Cambridge, 1991)
3. R.J. Aumann, B. Peleg, Von Neumann–Morgenstern solutions to cooperative games without side payments. Bull. Am. Math. Soc. **6**(3), 173–179 (1960)
4. W. Saad, Z. Han, M. Debbah, A. Hojrungnes, T. Basar, Coalition game theory for communication networks. IEEE Signal Process. Mag. **26**(5), 77–97 (2009)
5. Z. Han, D. Niyato, W. Saad, T. Basar, A. Hjorungnes, *Game Theory in Wireless and Communication Networks: Theory, Models and Applications* (Cambridge University Press, Cambridge, 2011)
6. D. Gillies, *Some Theorems on n-Person Games*. Ph.D. thesis, Department of Mathematics (Princeton University, Princeton, 1953)
7. G. Owen, *Game Theory*, 3rd edn. (Academic, London, 1995)
8. T. Sandholm, K. Larson, M. Anderson, O. Shehory, F. Tohme, Coalition structure generation with worst case guarantees. Artif. Intell. **10**, 200–238 (1999)
9. D. Ray, *A Game-Theoretic Perspective on Coalition Formation* (Oxford University Press, New York, 2007)
10. K. Apt, A. Witzel, A generic approach to coalition formation, in *Proceedings of the International Workshop Computational Social Choice (COMSOC)*, Amsterdam, The Netherlands (2006)
11. T. Arnold, U. Schwalbe, Dynamic coalition formation and the core. *J. Econ. Behav. Org.* **49**, 363–380 (2002)
12. W. Saad, Z. Han, M. Debbah, A. Hjorungnes, A distributed merge and split algorithm for fair cooperation in wireless networks, in *Proceedings of International Conferernce on Communications, Workshop Cooperative Communications and Networking*, Beijing, China (2008)
13. G. Chalkiadakis, E. Elkind, E. Markakis, N. R. Jennings, Cooperative games with overlapping coalitions. J. Artif. Intell. Res. **39**(1), 179–216 (2010)
14. Y. Zick, E. Elkind, Arbitrators in overlapping coalition formation games, in *Proceedings of 10th International Conference on Autonomous Agents and Multiagent Systems*, Taipei, Taiwan (2011)
15. Y. Zick, G. Chalkiadakis, E. Elkind, Overlapping coalition formation games charting the tractability frontier, in *Proceedings of 10th International Conference on Autonomous Agents and Multiagent Systems*, Valencia, Spain (2012)
16. T. Wang, L. Song, Z. Han, W. Saad, Overlapping coalition formation games for emerging communication networks. IEEE Netw. Mag. to appear (2016)

Chapter 2
Interference Management in Heterogenous Networks

2.1 Introduction

Small cell networks are seen as one of the most promising solutions for boosting the capacity and coverage of wireless networks. The basic idea of small cell networks is to deploy small cells, that are serviced by plug-and-play, low-cost, low-power small cell base stations (SBSs) able to connect to existing backhaul technologies (e.g., digital subscription line (DSL), cable modem, or a wireless backhaul) [1]. Types of small cells include operator-deployed picocells as well as femtocells that can be installed by end users at home or at the office. Recently, small cell networks have received significant attention from a number of standardization bodies including 3GPP [1, 2]. The deployment of SBSs is expected to deliver high capacity wireless access and enable new services for the mobile users while reducing the cost of deployment on the operators. Moreover, small cell networks are seen as a key enabler for offloading data traffic from the main, macrocellular network [3].

The successful introduction of small cell networks is contingent on meeting several key technical challenges, particularly, in terms of efficient interference management and distributed resource allocation [3–5]. For instance, underlying SBSs over the existing macrocellular networks leads to both cross-tier interference between the macrocell base stations and the SBSs and co-tier interference between small cells. If not properly managed, this increased interference can consequently affect the overall capacity of the two-tier network. There are two types of spectrum allocation for the network operator to select. The first type is orthogonal spectrum allocation, in which the spectrum in the network is shared in an orthogonal way between the macrocell and the small cell tiers. Although cross-tier interference can be totally eliminated using orthogonal spectrum allocation, the associated spectrum utilization is often inefficient [3]. The second type is co-channel assignment, in which both the macrocell and the small cell tiers share the same spectrum [4]. As the spectrum in the network is reused through co-channel assignment, the spectrum efficiency can be improved compared to the case of orthogonal spectrum allocation. However, both

© The Author(s) 2017
T. Wang et al., *Overlapping Coalition Formation Games
in Wireless Communication Networks*, SpringerBriefs in Electrical
and Computer Engineering, DOI 10.1007/978-3-319-25700-6_2

cross-tier interference and co-tier interference should be considered in this case. A lot of recent work has studied the problem of distributed resource allocation and interference management in small cells. These existing approaches include power control [6], fractional frequency reuse [7], interference alignment [8], interference coordination [9], the use of cognitive base stations [10], and interference cancellation [11].

Most existing works have focused on distributed interference management schemes in which the SBSs act noncooperatively. In such a noncooperative case, each SBS accounts only for its own quality of service while ignoring the co-tier interference it generates at other SBSs. Here, the co-tier interference between small cells becomes a serious problem that can significantly reduce the system throughput, particularly in outdoor picocell deployments. To overcome this issue, we enable cooperation between SBSs so as to perform cooperative interference management. The idea of cooperation in small cell networks has only been studied in a limited number of existing work [12–16]. In [12], the authors propose a cooperative resource allocation algorithm on intercell fairness in OFDMA femtocell networks. In [13], an opportunistic cooperation approach that allows femtocell users and macrocell users to cooperate is investigated. In [14], the authors introduce a game-theoretic approach to deal with the resource allocation problem of the femtocell users. In [15], a collaborative inter-site carrier aggregation mechanism is proposed to improve spectrum efficiency in a LTE-Advanced heterogeneous network with orthogonal spectrum allocation between the macrocell and the small cell tiers. The work in [16] propose a cooperative model for femtocell spectrum sharing using a cooperative game with transferable utility in partition form. However, the authors assume that the formed coalitions are disjoint and not allowed to overlap, which implies that each SBS can only join one coalition at most. This restriction on the cooperative abilities of the SBSs limits the rate gains from cooperation that can be achieved by the SBSs. Moreover, the authors in [16] adopt the approach of orthogonal spectrum allocation that is inefficient on spectrum occupation for the two-tier small cell networks.

The goal of this chapter is to develop a cooperative interference management model for small cell networks in which the SBSs are able to participate and cooperate with multiple coalitions depending on the associated benefit-cost tradeoff. We adopt the approach of co-channel assignment that improves the spectrum efficiency compared to the approach of orthogonal spectrum allocation used in [16]. We formulate the SBSs cooperation problem as an overlapping coalitional game and we present a distributed, self-organizing algorithm for performing overlapping coalition formation. Using the presented algorithm, the SBSs can interact and individually decide on which coalitions to participate in and on how much resources to use for cooperation. We show that, as opposed to existing coalitional game models that assume disjoint coalitions, this approach enables a higher flexibility in cooperation. We study the properties of this algorithm, and we show that it enables the SBSs to cooperate and self-organizing into the most beneficial and stable coalitional structure with overlapping coalitions. Simulation results show that this approach yields performance gains relative to both the noncooperative case and the classical case of coalitional games with nonoverlapping coalitions such as in [16].

2.2 System Model

Consider the downlink transmission of an orthogonal frequency division multiple access (OFDMA) small cell network composed of N SBSs and a macrocellular network having a single macro base station (MBS). The access method of all small cells and the macrocell is closed access. Let $\mathcal{N} = \{1, \ldots, N\}$ denote the set of all SBSs in the network. The MBS serves W macrocell user equipments (MUEs), and each SBS $i \in \mathcal{N}$ serves L_i small cell user equipments (SUEs). Let $\mathcal{L}_i = \{1, \ldots, L_i\}$ denote the set of SUEs served by an SBS $i \in \mathcal{N}$. Here, SBSs are connected with each other via a wireless backhaul. Each SBS $i \in \mathcal{N}$ chooses a subchannel set $\mathcal{T}_i$ containing $|\mathcal{T}_i| = M$ orthogonal frequency subchannels from a total set of subchannels $\mathcal{T}$ in a frequency division duplexing (FDD) access mode. The subchannel set $\mathcal{T}_i$ serves as the initial frequency resource of SBS $i \in \mathcal{N}$. The MBS also transmits its signal on the subchannel set $\mathcal{T}$, thus causing cross-tier interference from MBS to the SUEs served by the SBSs. Moreover, the SBSs are deployed in hot spot indoor large areas such as enterprises where there are no walls not only between each SBS and its associated SUEs, but also between all the SBSs. Meanwhile, the MBS is located outdoor, so there exist walls between the MBS and the SBSs.

In a traditional noncooperative scenario, each SBS $i \in \mathcal{N}$ transmits on its own subchannels. The set of the subchannels that SBS i owns is denoted as $\mathcal{T}_i$, where $\mathcal{T}_i \subseteq \mathcal{T}$. SBS i occupies the whole time duration of any subchannel $k \in \mathcal{T}_i$. Meanwhile, the MBS transmits its signal to the MUEs on several subchannels from $\mathcal{T}$, with each MUE occupying one subchannel at each time slot. When the SBSs act noncooperatively, each SBS uses all the subchannels from $\mathcal{T}_i$ to serve its SUEs $\mathcal{L}_i$. For each subchannel $k \in \mathcal{T}_i$, only one SUE $u \in \mathcal{L}_i$ is served on subchannel k. SUE u has access to the full time duration of subchannel k. We denote the channel gain between transmitter j and the receiver u that owns subchannel k in SBS i by g^k_{j,i_u} and the downlink transmit power from transmitter j and the receiver u that occupies subchannel k in SBS i by P^k_{j,i_u}. The rate of SBS $i \in \mathcal{N}$ in the noncooperative case is thus given by

$$\Upsilon_i = \sum_{k \in \mathcal{T}_i} \sum_{u \in \mathcal{L}_i} \log_2 \left(1 + \frac{P^k_{j,i_u} g^k_{j,i_u}}{\sigma^2 + I_{MN} + I_{SN}} \right), \tag{2.1}$$

where σ^2 is the variance of the Gaussian noise, $I_{MN} = P^k_{w,i_u} g^k_{w,i_u}$ is the cross-tier interference from the MBS w to a SUE served by SBS i on subchannel k, and $I_{SN} = \sum_{j \in \mathcal{N}, j \neq i} P^k_{j,i_u} g^k_{j,i_u}$ is the overall co-tier interference suffered by SUE u that is served by SBS i on subchannel k.

We note that, in dense small cell deployments, the co-tier interference between small cells can be extremely severe which can significantly reduce the rates achieved by the SBSs. Nevertheless, due to the wall loss and the long distance between MBS and SUEs, the downlink cross-tier interference is rather weak compared to the co-tier interference between small cells. Thus, in this work, we mainly deal with the downlink co-tier interference suffered by the SUEs from the neighboring SBSs. In order to deal with this interference problem, we allow the SBSs are to cooperate with

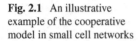

Fig. 2.1 An illustrative
example of the cooperative
model in small cell networks

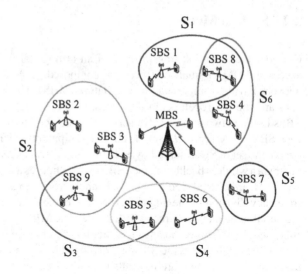

one another as illustrated in Fig. 2.1. In such a cooperative network, the SBSs can
cooperate to improve their performance and reduce co-tier interference.

2.3 Interference Management as OCF Games

2.3.1 K-Coalition OCF Game Model

Depending on signal-to-noise-plus-interference-ratio (SINR) feedbacks from their
SUEs, the SBSs can decide to form cooperative groups called coalitions so as to
mitigate the co-tier interference between neighboring SBSs within a coalition. The
SBSs can be modeled as the players in a coalitional game. Due to the fact that an
SBS may participate in multiple coalitions simultaneously as shown in Fig. 2.1, we
consider an OCF game model [17].

The SBSs in the network act as players joining coalitions. A coalition $\mathcal{R} = (\mathbf{R}_1, \ldots, \mathbf{R}_N)$ is a vector in which $\mathbf{R}_i$ is the subset of player i s resource set distributed
to this coalition. The support of a coalition $\mathcal{R}$ is defined as $\mathcal{C}(\mathcal{R}) = \{i \in \mathcal{N} | \mathbf{R}_i \neq \emptyset\}$.
After joining a coalition $\mathcal{R}$, SBS $i \in \mathcal{C}(\mathcal{R})$ allocates part of its frequency resource
into this coalition $\mathcal{R}$. Within each coalition $\mathcal{R}$, the SBSs can jointly coordinate their
transmission so as to avoid the collisions. The resource pool of coalition $\mathcal{R}$ is defined
as $\mathcal{T}_{\mathcal{R}} = \cup_{i \in \mathcal{C}(\mathcal{R})} \mathbf{R}_i$.

Without loss of generality, we consider that, whenever a coalition $\mathcal{R}$ successfully
forms, the transmissions inside $\mathcal{R}$ will be managed by a local scheduler using the
time division multiple access (TDMA) approach. The subchannels in $\mathcal{T}_{\mathcal{R}}$ are divided
into several time-slots. Each SBS can access only a fraction of all the time-slots when
transmitting on a specific subchannel. By doing so, the whole superframe duration

of each subchannel can be shared by more than one SBS. Hence, the downlink transmissions from each SBS in the coalition to its SUEs are separated. Consequently, no more than one SBS will be using the same subchannel on the same time-slot within a coalition, thus efficiently mitigating the interference inside the coalition $\mathcal{R}$. However, as the resource pools of different coalitions may not be disjoint, the system can still suffer from inter-coalition interference. Here, we note that this approach is still applicable under any other coalition-level interference mitigation scheme.

Specifically, we assume the resource pool $\mathcal{T}_{\mathcal{R}}$ of a coalition $\mathcal{R}$ is divided among the SBSs in $\mathcal{R}$ using a popular criterion named proportional fairness, i.e., each SBS $i \in C(\mathcal{R})$ gets an share $f_i \in [0, 1]$ of the frequency resources from the coalition $\mathcal{R}$ through the TDMA scheduling process of the local scheduler, and the share satisfies that $\sum_{i \in C(\mathcal{R})} f_i = 1$ and $f_i/f_j = |\mathbf{R}_i|/|\mathbf{R}_j|$. The proportional fairness criterion guarantees that the SBSs that dedicate more of its own frequency resources, i.e., subchannels to the coalition deserve more frequency resources back from the resource pool of the coalition. The gain of any coalition $\mathcal{R} \in CS$, which corresponds to the sum rate achieved by $\mathcal{R}$, is dependent on not only the members of $\mathcal{R}$ but also the coalitional structure CS due to inter-coalition interference. Formally, we define

$$U(\mathcal{R}, CS) = \sum_{i \in C(\mathcal{R})} \sum_{k \in T_i} \sum_{u \in \mathcal{L}_i} \gamma_{i,i_u}^k \log_2 \left(1 + \frac{P_{i,i_u}^k g_{i,i_u}^k}{\sigma^2 + I_{MO} + I_{SO}}\right), \qquad (2.2)$$

where γ_{i,i_u}^k denotes the fraction of the time duration during which SBS i transmits on channel k to serve SUE u, $I_{MO} = P_{w,i_u}^k g_{w,i_u}^k$ denotes the cross-tier interference from the MBS w to SUE u served by SBS i on subchannel k and $I_{SO} = \sum_{\mathcal{R}' \in CS, \mathcal{R}' \neq \mathcal{R}} \sum_{j \in C(\mathcal{R}), j \neq i} P_{j,i_u}^k g_{w,i_u}^j$ denotes the overall co-tier interference suffered by SUE u that is served by SBS i on subchannel k.

While cooperation can lead to significant performance benefits, it is also often accompanied by inherent coordination costs. In particular, for the considered SBS cooperation model, we capture the cost of forming coalitions via the amount of transmit power needed to exchange information. In each coalition $\mathcal{R}$, each SBS $i \in C(\mathcal{R})$ broadcasts its data to the other SBSs in the coalition. Here, each SBS needs to transmit the information to the farthest SBS in the same coalition. We assume that, during information exchange, no transmission errors occur. So the power cost incurred for forming a coalition $\mathcal{R}$ is $P_{\mathcal{R}} = \sum_{i \in C(\mathcal{R})} P_{i,j^*}$, where where P_{i,j^*} is the power spent by SBS i to broadcast the information to the farthest SBS j in a coalition $\mathcal{R}$. Meanwhile, for every coalition $\mathcal{R}$, we define a maximum tolerable power cost P_{lim}. Therefore, we define the value function of a coalition $\mathcal{R}$ as follows:

$$v(\mathcal{R}, CS) = \begin{cases} U(\mathcal{R}, CS), & \text{if } P_{\mathcal{R}} \leq P_{lim}, \\ 0, & \text{otherwise,} \end{cases} \qquad (2.3)$$

Therefore, the payoff that each SBS i achieves from coalition $\mathcal{R}$ is $p_i(\mathcal{R}, CS) = f_i v(\mathcal{R}, CS)$, and the total payoff of SBS i is then $x_i(CS) = \sum_{\mathcal{R} \in CS} p_i(\mathcal{R}, CS)$.

Table 2.1 Algorithm for Interference Management in HetNets

Initial State: The network consists of noncooperative SBSs, and the initial coalitional structure is denoted as $CS = \{\{T_1\}, \ldots, \{T_N\}\}$.

* **repeat**

1. Each SBS i discovers its nearby coalitions in the current coalition structure CS, the set of which is denoted by $\mathcal{N}_i$

2. SBS i finds a feasible transformation from the current coalition structure CS to a new coalition structure CS' by reallocating its resources among coalitions in $\mathcal{N}_i$, such that $x_i(CS) < x_i(CS')$

3. SBS i reallocate its resources and the network transforms to a new coalition structure CS'

* **until** the network converges to a stable coalition structure CS^*

2.3.2 Coalition Formation Algorithm

We note that the interference management in heterogenous networks is modeled as a K-coalition OCF game, where K is limited by the number of SBSs and the network topology. Based on the framework in Table 1.2, we give a distributed OCF algorithm to achieve a stable outcome of the K-coalition OCF game.

Definition 4 Given a set of player $S \subseteq \mathcal{N}$, we define a complete, reflexive, and transitive binary relation $\preceq_S$ on S over the set of all coalition structures, such that $CS_P \preceq_S CS_Q$, if and only if, we have $x_i(CS_P) \leq x_i(CS_Q)$ for any player $i \in S$.

Therefore, CS_Q is preferred to CS_P for the set of players S if no player's payoff is decreased by transforming CS_P to CS_Q. Based on these relationship, we can define a transform operation for the network. Formally, if the set of players S can transform the coalition structure from CS_P to CS_Q by reallocating their resources, and CS_Q is preferred to CS_P by the same set of players S, then, there exists a feasible transform from CS_P to CS_Q. Due to the communication cost, we restrict the set S as a single SBS and give a coalition formation algorithm as in Table 2.1.

This iterative algorithm starts from an initial state where each SBS forms a single coalition by devoting its own resources. Then, at each iteration, SBS i discovers the nearby coalitions through environment sensing [18], finds a feasible and profitable transform by using the defined relationship $\preceq_{\{i\}}$ and reallocates its resources to perform the transformation. The network converges when there is no feasible and profitable transform for any SBS, and outputs a stable coalition structure CS^*. Given the stable coalition structure CS^*, each SBS i devotes the corresponding resources $\mathbf{R}_i$ to each coalition $\mathcal{R} \in CS^*$. For each coalition $\mathcal{R} \in CS^*$, the coalition members $\mathcal{C}(\mathcal{R})$ coordinate with each other by rescheduling their transmissions with TDMA using the resource pool $T_{\mathcal{R}}$.

2.3.3 Simulation Results

For simulations, we consider an MBS that is located at the chosen coordinate of (1 km, 1 km). The radius of the coverage area of the MBS is 0.75 km. The number of MUEs is 10. N SBSs are deployed randomly and uniformly within a circular area around the MBS with a radius of 0.1 km. There is a wall loss attenuation of 20 dB between the MBS and the SUEs, and no wall loss between the SBSs and the SUEs. Each SBS has a circular coverage area with a radius of 20 m. Each SBS has 4 subchannels to use and serves 4 users as is typical for small cells. The total number of subchannels in the considered OFDMA small cell network is 20. The bandwidth of each subchannel is 180 kHz. The total number of time-slots in each transmission in TDMA mode is 4. The transmit power of each SBS is set at 20 dBm, while the transmit power of the MBS is 35 dBm. The maximum tolerable power to form a coalition $P_{lim} = 100$ dBm. The noise variance is -104 dBm.

In Fig. 2.2, we present a snapshot of an OFDMA small cell network resulting from the given algorithm with $N = 7$ SBSs. The radius of the distribution area of SBSs is 0.7 km. The cooperative network shown in this figure is a stable coalitional structure CS^*. Initially, all the SBSs schedule their transmissions noncooperatively. After using the OCF algorithm, they self-organize into the structure in Fig. 2.2. This coalitional structure consists of 5 overlapping coalitions named Coalition 1, Coalition 2, Coalition 3, Coalition 4, and Coalition 5. The support of Coalition 1 consists of SBS 3 and SBS 6. The support of Coalition 2 includes SBS 2 and SBS 5. The support of Coalition 3 includes SBS 1 and SBS 6. The support of Coalition 4 includes SBS

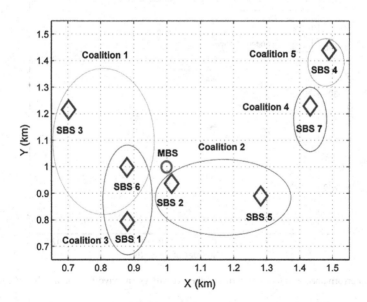

Fig. 2.2 A snapshot of an overlapping coalitional structure resulting from the considered approach in a small cell network

7. The support of Coalition 5 includes SBS 4. SBS 4 and SBS 7 have no incentive to cooperate with other SBSs as their spectral occupation is orthogonal to all nearby coalitions. Meanwhile, SBS 6 is an overlapping player because its resource units are divided into two parts assigned to different coalitions. The interference is significantly reduced in CS^* as compared to that in the noncooperative case, as the interference between the members of the same coalition is eliminated using proper scheduling. Clearly, Fig. 2.2 shows that by adopting the OCF algorithm, the SBSs can self-organize to reach the final network structure.

Figure 2.3 shows the overall system utility in terms of the total rate achieved by the OCF algorithm as a function of the number of SBSs N compared with two other cases: the nonoverlapping coalition formation (CF) algorithm and the noncooperative case. Figure 2.3 shows that for small networks ($N < 4$), due to the limited choice for cooperation, the OCF algorithm and the CF algorithm have a performance that is only slightly better than that of the noncooperative case. This indicates that the SBSs have no incentive to cooperate in a small-sized network as the co-tier interference remains tolerable and the cooperation possibilities are small. As the number of SBS N increases, the possibility of cooperation for mitigating interference increases. Figure 2.3 shows that, as N increases, the OCF algorithm exhibits improved system performances compared to both the traditional coalition formation game and that of the noncooperative case. The performance advantage reaches up to 32 and 9 % at $N = 10$ SBSs relative to the noncooperative case and the classical CF case, respectively.

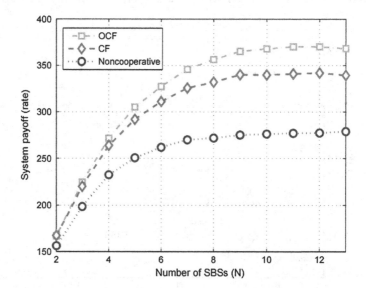

Fig. 2.3 Performance evaluation in terms of the overall system payoff as the number of SBSs N varies

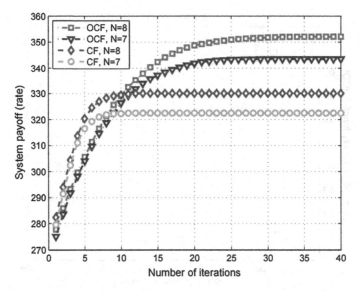

Fig. 2.4 System payoff versus number of iterations

Figure 2.4 shows the convergence process under different scenarios using the OCF algorithm and the CF algorithm. We observe that, although the OCF algorithm requires a few additional iterations to reach the convergence as opposed to the CF case when both $N = 7$ and $N = 8$, this number of iterations for OCF remains reasonable. Moreover, Fig. 2.4 shows that the OCF algorithm clearly yields a higher system payoff than the CF case, with only little extra overhead, in terms of the number of iterations. Hence, the simulation results in Fig. 2.4 clearly corroborate our earlier analysis.

Figure 2.5 shows the cumulative density function (CDF) of the individual SBS payoff resulting from the OCF algorithm and the CF algorithm when the number of SBSs is set to $N = 10$. From Fig. 2.5, we can clearly see that the OCF algorithm performs better than the CF algorithm in terms of the individual payoff per SBS. For example, the expected value of the individual payoff for a network formed from the OCF algorithm is 36, while for a network formed from the CF algorithm the expected value is 33. This is due to that the OCF algorithm allows more flexibility for the SBSs to cooperate and form coalitions. Each SBS is able to join multiple coalitions in a distributed. way by adopting our OCF algorithm, while it can only join one coalition at most in the CF case. Moreover, during each reallocation, the SBSs improve their own payoff without being detrimental to the other SBSs in the new coalition. This also contribute to a growth of the individual payoff of each SBS. In a nutshell, Fig. 2.5 shows that the OCF algorithm yields an advantage on individual payoff per SBS over the CF algorithm.

Figure 2.6 shows the growth of the system payoff of the network as the number of SBSs increases, under different maximum tolerable power costs of a coalition P_{lim}.

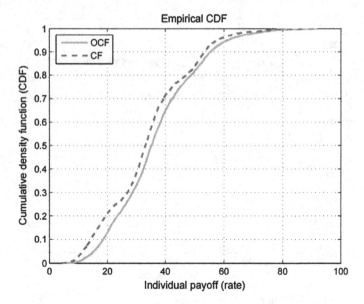

Fig. 2.5 Cumulative density function of the individual payoff for a network with $N = 10$ SBSs

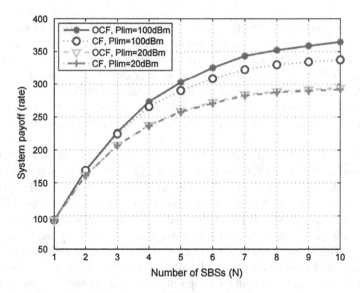

Fig. 2.6 System payoff as a function of number of SBSs N, for different maximum tolerable power costs

Both the OCF algorithm and the CF case are considered in Fig. 2.6. We observe that, as the number of SBSs increases, the system payoff under two conditions both grows. Moreover, the OCF algorithm has a small advantage on the system payoff compared to the CF case when $P_{lim} = 20$ dBm, while the advantage of the OCF algorithm over the CF case is more significant when $P_{lim} = 100$ dBm. This is due to the fact that when P_{lim} is low, the SBSs can hardly cooperate with other neighboring SBSs. Most SBSs choose to stay alone as the power cost of possible coalitions exceeds the maximum tolerable power cost. Thus, the system payoff of the OCF algorithm and of the CF algorithm are close. Furthermore, when P_{lim} is high, each SBS is able to reallocate its SBS units to join neighboring coalitions and improve both the system payoff and its own payoff using the OCF algorithm. Meanwhile, the cooperation possibility of the SBSs under the CF case is also increased when P_{lim} increases. Consequently, Fig. 2.6 shows that the OCF algorithm incurs a higher probability for the SBSs to cooperate than the CF case, especially when the maximum tolerable power cost of forming a coalition is high. Thus, our OCF algorithm achieves better system performances in terms of sum rate than the CF algorithm.

Figure 2.7 shows the relationship between the number of coalitions that each SBS joins and the number of SBSs under the OCF case and the CF case. As the number of SBSs increases, both the maximum and the average number of coalitions that each SBS joins also grows under the OCF case. While in the CF case, each SBS is only allowed to join one coalition at most no matter how the number of SBSs changes, thus causing the maximum number and the average number of coalitions that each SBS joins to remain the same when the number of SBSs increases. Figure 2.7 shows that the incentive toward cooperation for the SBSs is more significant for the

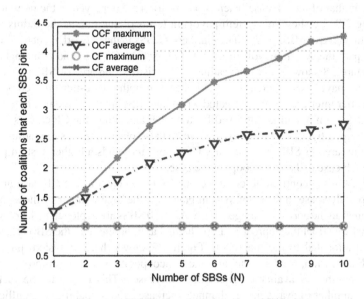

Fig. 2.7 Number of coalitions per SBS as a function of number of SBSs N

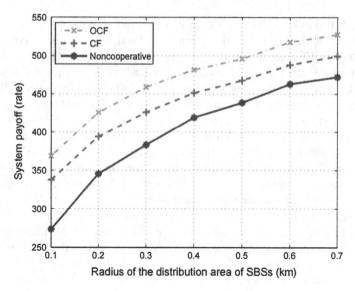

Fig. 2.8 System payoff versus radius of the distribution area of SBSs for a network with $N = 10$ SBSs

OCF algorithm than for the CF case. Thus, The cooperative gain can be achieved more efficiently by using our OCF algorithm than the CF case when the SBSs are densely deployed in the network. The cooperative probability of the OCF algorithm represented by the maximum number of coalitions that each SBS joins is 325.75 % larger than that of the CF case when $N = 10$ SBSs are deployed in the network.

In Fig. 2.8, we show the system payoff in terms of sumrate as the radius of the distribution area of SBSs varies. The number of SBSs in the network is set to $N = 10$. We compare the system payoff of the OCF algorithm, CF case and noncooperative case. Figure 2.8 shows that as the radius of the distribution area of SBSs increases, the system payoff also increases. This is because both the co-tier interference and the cross-tier interference are mitigated when the SBSs are deployed in a larger area. Thus, the system payoff is improved for the OCF algorithm, the CF case as well as the noncooperative case. From Fig. 2.8, we can also observe that as the radius of the distribution area of SBSs varies, our OCF algorithm yields a higher system payoff than the CF case and the noncooperative case.

In Fig. 2.9, we continue to compare our OCF approach to the CF case and the noncooperative case in terms of system payoff as the total number of the available subchannels in the network changes. Here, $N = 10$ SBSs are deployed in the network. Note that, we adopt the approach of co-channel assignment, i.e., the SBSs reuse the spectrum allocated to the macrocell. Figure 2.9 shows that the system payoff of the OCF algorithm, the CF case, and the noncooperative case are improved when the total number of available subchannels increases. This is due to the fact that when the number of available subchannels increases, the probability of conflicts on subchannels is greatly decreased. Thus, the interference in the two-tier small cell

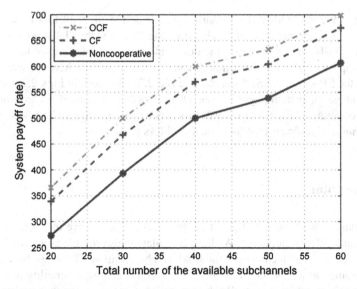

Fig. 2.9   System payoff versus total number of subchannels for a network with $N = 10$ SBSs

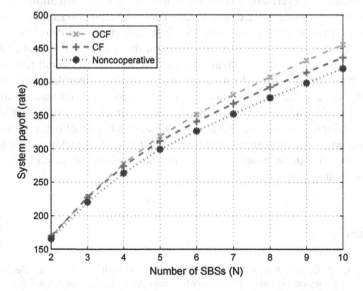

Fig. 2.10   Performance evaluation in terms of the overall system payoff with wall loss in the small cell tier as the number of SBSs N varies

network is mitigated, causing the improvement of the system payoff in terms of sum rate. Moreover, Fig. 2.9 shows that the OCF algorithm outperforms the CF case and the noncooperative case in terms of system payoff when the total number of available subchannels increases.

In Fig. 2.10, we modify the scenario by considering the wall loss between the MBS and the SUEs and the wall loss between the SBSs and the SUEs, both of which are set at 20 dB. In this scenario, the downlink cross-tier interference has a much greater impact on system performance than in the scenario where no wall exists between the SBSs and the SUEs such as in Fig. 2.3. As shown in Figs. 2.3 and 2.10, the advantage on system payoff of our OCF algorithm over the CF algorithm and the noncooperative case when no wall loss is considered between the SBSs and the SUEs is more significant than that when wall loss is involved.

2.4 Summary

In this chapter, we have investigated the problem of cooperative interference management in small cell networks. We have formulated this problem as an overlapping coalition formation game between the small cell base stations. Then, we have shown that the game has a transferable utility and exhibits negative externality due to the co-tier interference between small cell base stations. To solve this game, we have presented a distributed overlapping coalition formation algorithm that allows the small cell base stations to interact and individually decide on their cooperative decisions. By adopting this algorithm, each small cell base station can decide on the number of coalitions that it wishes to join as well as on the resources that it allocates to each such coalition, while optimizing the tradeoff between its overall rate and the associated cooperative costs. We have shown that the OCF algorithm is guaranteed to converge to a stable coalition structure in which no small cell base station has an incentive to reallocate its cooperative resources. Simulation results have shown that the overlapping coalitional game approach allows the small cell base stations to self-organize into cooperative coalitional structures while yielding notable rate gains relative to both the noncooperative case and the classical coalition formation algorithm with nonoverlapping coalitions.

References

1. T.Q.S. Quek, G. de la Roche, I. Guvenc, M. Kountouris, *Femtocell Networks: Deployment, PHY Techniques, and Resource Management*. (Cambridge University Press, Cambridge, 2013)
2. J.G. Andrews, H. Claussen, M. Dohler, S. Rangan, M. Reed, Femtocells: Past, Present, and Future. IEEE J. Sel. Areas Commun. **30**(3), 497–508 (2012)
3. D. Lopez-Perez, A. Valcarce, G. de la Roche, J. Zhang, OFDMA Femtocells: a roadmap on interference avoidance. IEEE Commun. Mag. **47**(9), 41–48 (2009)
4. D. Calin, H. Claussen, H. Uzunalioglu, On femto deployment architectures and macrocell offloading benefits in joint macro-femto deployments. IEEE Commun. Mag. **48**(1), 26–32 (2010)
5. Y.S. Liang, W.H. Chung, G.K. Ni, I.Y. Chen, H. Zhang, S.Y. Kuo, Resource allocation with interference avoidance in OFDMA femtocell networks. IEEE Trans. Veh. Technol. **61**(5), 2243–2255 (2012)

6. V. Chandrasekhar, J. Andrews, Z. Shen, T. Muharemovic, A. Gatherer, Power control in two-tier femtocell networks. IEEE Trans. Wirel. Commun. **8**(8), 4316–4328 (2009)
7. J.Y. Lee, S.J. Bae, Y.M. Kwon, Interference analysis for femtocell deployment in OFDMA systems based on fractional frequency reuse. IEEE Commun. Lett. **15**(4), 425–427 (2011)
8. N. Lertwiram, P. Popovski, K. Sakaguchi, A study of trade-off between opportunistic resource allocation and interference alignment in femtocell scenarios. IEEE Commun. Lett. **1**(4), 356–359 (2012)
9. S. Randan, R. Madan, Belief propagation methods for intercell interference coordination in femtocell networks. IEEE J. Sel. Areas Commun. **30**(3), 631–640 (2012)
10. J. W. Huang, V. Krishnamurthy, Cognitive base stations in LTE/3GPP femtocells: a correlated equilibrium game-theoretic approach. IEEE Trans. Commun. **59**(12), 3485–3493 (2011)
11. M. Wildemeersch, T.Q.S. Quek, M. Kountouris, C.H. Slump, Successive interference cancellation in uplink cellular networks, in *Proceedings 2013 IEEE SPAWC*, Darmstadt, Germany (2013)
12. K. Lee, O. Jo, D.H. Cho, Cooperative resource allocation for guaranteeing intercell fairness in femtocell networks. IEEE Commun. Lett. **15**(2), 214–216 (2011)
13. R. Urgaonkar, M.J. Neely, Opportunistic cooperation in cognitive femtocell networks. IEEE Trans. Veh. Technol. **30**(3), 607–616 (2012)
14. O.N. Gharehshiran, A. Attar, V. Krishnamurthy, Collaborative sub-channel allocation in cognitive LTE femto-cells: a cooperative game theoretic approach.IEEE Trans. Commun. **61**(1), 325–334 (2013)
15. B. Soret, H. Wang, K.I. Pedersen, C. Rosa, Multicell cooperation for LTE-advanced heterogeneous network scenarios. IEEE Wireless Commun. **20**(1), 27–34 (2013)
16. F. Pantisano, M. Bennis, W. Saad, R. Verdone, M. Latva-aho, Coalition formation games for femtocell interference management: a recursive core approach, in *Proceedings of the IEEE Wireless Communication Network Conference*, Quintana-Roo, Mexico (2011)
17. Z. Zhang, L. Song, Z. Han, W. Saad, Coalitional games with overlapping coalitions for interference management in small cell networks. IEEE Trans. Wireless Commun. **12**(5), 2659–2669 (2014)
18. A. Zalonis, N. Dimitriou, A. Polydoros, J. Nasreddine, P. Mahonen, Femtocell downlink power control based on radio environment maps, in *Proceedings of the IEEE Wireless Communication Network Conference*, Paris, France (2012)

Chapter 3
Cooperative Spectrum Sensing in Cognitive Radio

3.1 Introduction

Cognitive radio (CR) has been proposed to increase spectrum efficiency, in which unlicensed, secondary users (SUs), can sense the environment and change their parameters to access the spectrum of licensed, primary users (PUs), while maintaining the interference to the PUs below a tolerable threshold [1]. In order to exploit the spectrum holes, the SUs must be able to smartly sense the spectrum so as to decide which portion can be exploited [2]. Depending on the features of different signals, different spectrum sensing detectors have been designed, such as energy detectors, waveform-based detectors, and matched-filtering detectors [3]. However, the performance of these detectors is highly susceptible to the noise, small-scale fading, and shadowing over wireless channels. To overcome this problem, cooperative spectrum sensing (CSS) was proposed, in which the SUs utilize the natural space diversity by sharing sensing results among each other and making collaborative decision on the detection of PUs [4–25]. It has been shown that CSS can significantly improve the sensing accuracy, in comparison with the conventional, noncooperative case which relies solely on local detectors.

According to [4], the CSS schemes can be classified into three categories based on how the sensing data is shared in the network: centralized [5–10], relay-assisted [11, 12], and distributed [13–17]. In centralized CSS, a common fusion center (FC) collects sensing data from all the SUs in the network via a reporting channel, then combines the received local sensing data to determine the presence or absence of PUs, and at last diffuses the decision back to the SUs. In relay-assisted CSS, there is also a common FC, but the local sensing data, instead of being transmitted directly to the FC, is relayed by the SUs so as to reduce transmission errors. Unlike the centralized or relay-assisted CSS, distributed cooperative sensing (DCS) does not rely on an FC

© The Author(s) 2017
T. Wang et al., *Overlapping Coalition Formation Games in Wireless Communication Networks*, SpringerBriefs in Electrical and Computer Engineering, DOI 10.1007/978-3-319-25700-6_3

for making the cooperative decision. In this case, each SU simultaneously sends and receives sensing data via the reporting channel, and then combines the received data using a local fusion rule. Therefore, the SUs in DCS can make individual decisions on whether to access the spectrum, and thus, can adapt to the situation in which the SUs belong to different authorities or operators and distributed decisions must be made. Hereinafter, we focus on DCS.

In [13], the authors propose a coalition-based DCS, in which the SUs self-organize into *disjoint* coalitions, and apply centralized CSS inside each coalition. The coalition formation process is based on a coalition formation game (CF game) with nontransferable utility [26, 27], which jointly considers the associated benefit and cost for forming coalitions. This coalition-based DCS, in which the signaling overhead is shared by the coalition heads that are much closer to the SUs, can largely decrease the bandwidth requirement for reporting local sensing results. Other approaches that studied DCS are found in [14–17]. However, in [13–17], the network structure is *restricted to disjoint, nonoverlapping coalitions*, which implies that the local sensing results of an SU can only be shared within a single coalition, although, for the coalition-edge SUs, their local sensing results can be efficiently transmitted to the nearby coalitions for further improving the cooperative sensing performance. Hence, this disjoint coalitional structure of SUs may limit the gains from DCS and, thus, to reap the gains of DCS, information sharing among multiple coalitions should be considered.

Traditionally, the SUs are assumed to share the same occupancy of PUs, i.e., whether the PU is present for all SUs or it is absent for all SUs. However, in practical systems, due to location and time diversities, the SUs may experience different spectrum occupancies. Some recent studies have noticed this problem and algorithms for spectrum-heterogeneous cognitive radio systems have been proposed [18–21]. Besides the diversity of SUs, other issues that greatly influence the sensing performance of SUs have also been studied, e.g., the spatial correlation between SUs [22, 23], the mobility of PUs [24], the nonidealness of the report channel [25].

The goal of this is to develop a DCS approach in which SUs can share their sensing information with a multitude of coalitions [28, 29]. In particular, we consider two criteria to evaluate the sensing performance, and for each criterion, we formulate the general DCS problem as an optimization with strict power and bandwidth constraints. In order to solve the DCS problem distributively, we introduce a new overlapping coalition formation (OCF) approach, which significantly differs from the existing nonoverlapping DCS such as in [13] as it allows each SU to cooperate with *multiple, overlapping coalitions* by allocating each coalition a portion of its local power and bandwidth resources. In particular, we introduce *overlapping coalition formation games* [30–32], to model the DCS problem, and we present a distributed algorithm that is shown to converge to a stable coalitional structure with overlapping coalitions. Simulation results show that the overlapping algorithm yields significant performance improvements compared with the state-of-the-art nonoverlapping algorithm for all network scenarios while also reducing the required overhead and system complexity.

3.2 System Model

Consider a cognitive radio network with N SUs equipped with energy detectors [7], the set of which is denoted by $\mathcal{N} = \{1, 2, \ldots, N\}$, and a single PU far away from them [7, 8]. The distance between SU i and SU j is denoted by $d_{i,j}$. The distance between the PU and any SU is denoted by D, and we have $D \gg d_{i,j}$ for any SU i and j. In this network, the SUs individually and locally decide on the presence or absence of the PU via their own local information. We assume that the SUs can cooperate with one another by exchanging their sensing data via a reporting channel, and the overall DCS phase consists of three successive periods: the local sensing period, the data reporting period and the data fusion period. In the local sensing period, each SU locally detects the presence of the PU on the sensing channel. In the data reporting period, each SU sends its own sensing data to other SUs via the reporting channel with power and bandwidth constraints. In the data fusion period, each SU combines its local sensing data with the received sensing data and decides whether or not the PU is present. Once the DCS phase is completed, each SU locally decides whether to access the spectrum based on its decision of the PU' state as well as the particular distributed protocol used at the MAC layer, such as the distributed MAC protocols in [33, 34]. Figure 3.1 illustrates the DCS process described above in a CR network with 3 SUs.

3.2.1 Local Sensing

We denote by $\mathcal{H}_1$ and $\mathcal{H}_0$ the hypotheses of the presence and absence of the PU, respectively. The sampled signal at SU $i \in \mathcal{N}$ is given by:

$$y_i(n) = \begin{cases} h_i(n)s(n) + u_i(n), & \mathcal{H}_1, \\ u_i(n), & \mathcal{H}_0, \end{cases} \tag{3.1}$$

where $h_i(n)$ denotes the channel between the PU and SU i, $s(n)$ denotes the signal from the PU and $u_i(n)$ denotes the noise at SU i. In accordance with [7], we assume $s(n)$ is an independent identically distributed (i.i.d.) random process with zero mean and variance σ_s^2, $u_i(n)$ is i.i.d. Gaussian with zero mean and variance σ_u^2, $|h_i(n)|$ is Rayleigh distributed. Since the distances between any SUs are negligible compared with the distance from the PU to any SU, $|h_i(n)|, i \in \mathcal{N}$ are assumed to have the same variance $\sigma_h^2 = \kappa D^{-\mu}$, where κ and μ are path loss parameters. For any SU

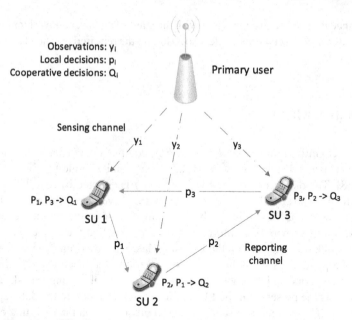

Fig. 3.1 Illustration of distributed cooperative sensing in a cognitive radio network with 3 SUs

$i \in \mathcal{N}$, the energy detector's probabilities of missed detection and false alarm are, respectively, given by [7]:

$$P_{m,i}(\lambda_i) = 1 - \mathcal{Q}\left(\left(\frac{\lambda_i}{1+\gamma} - 1\right)\sqrt{N_s}\right), \tag{3.2}$$

$$P_{f,i}(\lambda_i) = \mathcal{Q}\left((\lambda_i - 1)\sqrt{N_s}\right), \tag{3.3}$$

where $\mathcal{Q}(\cdot)$ denotes the right-tail probability of a normalized Gaussian distribution, $\gamma = \sigma_h^2 \sigma_s^2 / \sigma_u^2$ is the average received SNR at each SU, $\lambda_i \sigma_u^2$ is the threshold of the energy detector at SU i, and N_s is the product of the sensing time and sampling frequency. We assume γ and N_s are constant parameters.

3.2.2 Data Reporting

In order to reduce the bandwidth for reporting, the local sensing data is quantized to 1 bit (hard decisions) in [6]. In addition, we assume that each SU has limited transmit power P_{SU} and time-frequency resource θ_{SU} during the data reporting period. For the reporting between any two SUs, the minimum average received SNR is assumed to be γ_0 and the minimum time-frequency resource for transmitting 1 bit is assumed to be θ_0. Therefore, the power and bandwidth constraints for any SU $i \in \mathcal{N}$ are given by:

$$\sum_{j \in S_i} \frac{\gamma_0 \sigma_u^2}{\kappa d_{i,j}^{-\mu}} \leq P_{SU}, \tag{3.4}$$

$$|S_i| \theta_0 \leq \theta_{SU}, \tag{3.5}$$

where S_i is the "report-to" set of SU i consisting of the SUs that SU i reports to.

3.2.3 Data Fusion

After every SU sends the sensing data to its designated receivers, each SU combines all the received sensing data (including its local sensing data) using a local fusion rule. Suppose that the k_i-out-of-all fusion rule is adopted by SU i [7, 8], i.e., SU i decides the presence of the PU if at least k_i reports declare that the PU is detected, and vice versa. Consequently, SU i's probabilities of missed detection and false alarm are, respectively, given by:

$$Q_{m,i}(k_i) = \sum_{|\mathcal{R}_i^1| < k_i} \left[\prod_{j \in \mathcal{R}_i^1} (1 - P_{m,j}) \prod_{j \in \mathcal{R}_i^0} P_{m,j} \right], \tag{3.6}$$

$$Q_{f,i}(k_i) = \sum_{|\mathcal{R}_i^1| \geq k_i} \left[\prod_{j \in \mathcal{R}_i^1} P_{f,j} \prod_{j \in \mathcal{R}_i^0} (1 - P_{f,j}) \right], \tag{3.7}$$

where $\mathcal{R}_i^1 \cup \mathcal{R}_i^0 = \mathcal{R}_i$ is the "report-from" set of SU i which consists of SU i it as well as the SUs that report to SU i, and $\mathcal{R}_i^1$, $\mathcal{R}_i^0$ denote the set of SUs whose reports declare the presence and absence of the PU, respectively. Note that k_i is an integer between 1 and $|\mathcal{R}_i|$.

3.3 DCS as an Optimization Problem

From the system model, we can see that the DCS process is determined by local parameters as well as by the reporting structure of the network, i.e., the local sensing thresholds $\lambda_i, i \in \mathcal{N}$, the local fusion rules $k_i, i \in \mathcal{N}$, and the report-to sets $S_i, i \in \mathcal{N}$, or equally, the report-from sets $\mathcal{R}_i, i \in \mathcal{N}$. We consider a $1 \times N$ vector $\mathbf{\Lambda} = (\lambda_1, \lambda_2, \ldots, \lambda_N)$ as the local threshold vector, a $1 \times N$ vector $\mathbf{K} = (k_1, k_2, \ldots, k_N)$ as the fusion rule vector, and an $N \times N$ binary matrix $\mathbf{\Omega} = \{\omega_{i,j}\}, \omega_{i,j} = \{0, 1\}$ as the reporting matrix where $\omega_{i,j} = 1$ implies that SU i's report is received by SU j. Note that the "report-to" sets $\mathcal{S}_i$, $i \in \mathcal{N}$ and the "report-from" sets $\mathcal{R}_i$, $i \in \mathcal{N}$ are given by the rows and columns of $\boldsymbol{\Omega}$, respectively. To evaluate the performance of DCS, we consider two criteria that are commonly used in the literature, the $Q_m + Q_f$ criterion [8] and Q_m/Q_f criterion [13].

3.3.1 $Q_m + Q_f$ Criterion

In the $Q_m + Q_f$ criterion, we consider the probability that the cooperative sensing decision is incorrect, which is referred to as the "total error rate" in [8]. Strictly speaking, the total error rate of an SU $i \in \mathcal{N}$ is given by $P_1 Q_{m,i} + (1 - P_1) Q_{f,i}$, where P_1 is the probability that the PU is present. For conciseness, we assume $P_1 = 0.5$, and thus, the total error rate is given by $(Q_{m,i} + Q_{f,i})/2$. Moreover, we consider the average sensing performance of all SUs in the network, i.e., $(1/2N) \sum_{i \in \mathcal{N}} (Q_{m,i} + Q_{f,i})$. By omitting the factor $1/2N$ from the objective function, we have the DCS problem is formulated as:

$$\min_{\boldsymbol{\Lambda}, \mathbf{K}, \boldsymbol{\Omega}} \sum_{i \in \mathcal{N}} \left(Q_{m,i} + Q_{f,i} \right), \tag{3.8a}$$

$$s.t. \sum_{j \neq i, \omega_{i,j} = 1} \frac{\gamma_0 \sigma_u^2}{\kappa d_{i,j}^{-\mu}} \leq P_{SU}, \; i = 1, 2, \ldots, N, \tag{3.8b}$$

$$\sum_{j \neq i, \omega_{i,j} = 1} \theta_0 \leq \theta_{SU}, \; i = 1, 2, \ldots, N, \tag{3.8c}$$

where $Q_{m,i}$ and $Q_{f,i}$ are given by (3.6) and (3.7), respectively. We note that our model and analysis can be extended to the more general setting with any P_1, in a straightforward manner, and our results still hold.

Problem (3.8) is a mixed integer nonlinear programming problem that is known to be intractable in the general case [35]. Moreover, due to the lack of a fusion or control center in the considered network, any possible centralized algorithm that gives an optimal solution will not be applicable for the DCS process. Therefore, we consider suboptimal solutions with distributed algorithms. Note that the constraints (3.8b) and (3.8c) are only related to $\boldsymbol{\Omega}$; we consider a suboptimal solution with two separate steps:

(a) Find a feasible reporting matrix $\boldsymbol{\Omega}$ that satisfies the constraints in (3.8b) and (3.8c).
(b) Compute the optimal $\boldsymbol{\Lambda}$ and $\mathbf{K}$ for the objective function (3.8a) with $\boldsymbol{\Omega}$ given in step a.

To simplify the problem, we assume that the AND rule is adopted by all SUs, i.e., $k_i = |\mathcal{R}_i|, i \in \mathcal{N}$, and thus, step b is reduced to the computation of the optimal $\mathbf{\Lambda}$ for the objective function with the given $\mathbf{\Omega}$ and $\mathbf{K} = (|\mathcal{R}_1|, |\mathcal{R}_2|, \ldots, |\mathcal{R}_N|)$, where $|\mathcal{R}_i| = \sum_{j \in \mathcal{N}} \omega_{j,i}, i \in \mathcal{N}$. By substituting $\mathbf{\Omega}, \mathbf{K}$ and (3.2), (3.3) into (3.6), (3.7), and further substituting (3.6) and (3.7) into (3.8a), step b is formally written as:

$$\min_{\mathbf{\Lambda}} \sum_{i \in \mathcal{N}} \left[1 - \prod_{j \in \mathcal{N}, \omega_{j,i}=1} \mathcal{Q}\left(\left(\frac{\lambda_j}{1+\gamma} - 1\right)\sqrt{N_s}\right) \right.$$
$$\left. + \prod_{j \in \mathcal{N}, \omega_{j,i}=1} \mathcal{Q}\left((\lambda_j - 1)\sqrt{N_s}\right) \right]. \tag{3.9}$$

3.3.2 Q_m/Q_f Criterion

In the Q_m/Q_f criterion, the network sensing performance is evaluated via the average value of one error probability while the other probability is maintained below a certain threshold α. Here, we consider the average value of the missed detection probability while the false alarm probabilities are such that $Q_{f,i} \leq \alpha, i \in \mathcal{N}$. This criterion indicates the interference to the PU while we guarantee a usability rate of the spectrum holes. Mathematically, the DCS problem is formulated as:

$$\min_{\mathbf{\Lambda}, \mathbf{K}, \mathbf{\Omega}} \sum_{i \in \mathcal{N}} Q_{m,i}, \tag{3.10a}$$

$$s.t. \quad \sum_{j \neq i, \omega_{i,j}=1} \frac{\gamma_0 \sigma_u^2}{\kappa d_{i,j}^{-\mu}} \leq P_{SU}, \ i = 1, 2, \ldots, N, \tag{3.10b}$$

$$\sum_{j \neq i, \omega_{i,j}=1} \theta_0 \leq \theta_{SU}, \ i = 1, 2, \ldots, N, \tag{3.10c}$$

$$Q_{f,i} \leq \alpha, \ i = 1, 2, \ldots, N, \tag{3.10d}$$

where $Q_{m,i}$ and $Q_{f,i}$ are given by (3.6) and (3.7), respectively.

Problem (3.10) is also a mixed integer nonlinear programming problem. For similar reasons as in the $Q_m + Q_f$ criterion, we consider a suboptimal solution with two separate steps:

(a) Find a feasible reporting matrix $\mathbf{\Omega}$ that satisfies the constraints in (3.10b) and (3.10c).
(b) Compute the optimal $\mathbf{\Lambda}, \mathbf{K}$ for (3.10a) with $\mathbf{\Omega}$ given in step (a) and the constrains in (3.10d).

Note that step (a) is exactly the same as in the $Q_m + Q_f$ criterion. For step (b), we also assume the AND rule is adopted by all SUs and, thus, it reduces to:

$$\min_{\Lambda} \sum_{i \in \mathcal{N}} \left[1 - \prod_{j \in \mathcal{N}, \omega_{j,i}=1} \mathcal{Q}\left(\left(\frac{\lambda_j}{1+\gamma} - 1 \right) \sqrt{N_s} \right) \right] \tag{3.11a}$$

$$s.t. \prod_{j \in \mathcal{N}, \omega_{j,i}=1} \mathcal{Q}\left((\lambda_j - 1) \sqrt{N_s} \right) \leq \alpha, \ i = 1, \ldots, N. \tag{3.11b}$$

3.4 DCS Based on Overlapping Coalition Formation Games

In the previous section, the DCS problem is divided into two separate subproblems. The first subproblem aims to find a feasible reporting matrix $\mathbf{\Omega}$. The second subproblem aims at computing the optimal sensing threshold vector $\mathbf{\Lambda}$ with the given $\mathbf{\Omega}$. In this section, we consider the DCS problem as an OCF game, in which the first subproblem is strictly modeled as the local resource limitation, and the second subproblem is captured by an adequately designed utility function for each coalition of SUs. Based on the OCF-game model, we present a distributed coalition formation algorithm that allows to form overlapping coalitions, and a threshold decision algorithm that locally decides the sensing threshold of each SU. Note that the DCS problems in both the $Q_m + Q_f$ and Q_m/Q_f criteria are uniformly modeled by the OCF game. These algorithms apply to both criteria.

3.4.1 OCF-Game Model

In essence, coalitional games involve a set of players who seek to form cooperative groups, i.e., coalitions, to strengthen their positions in a given game scenario [26]. In particular, in an OCF game [30], the players can join multiple coalitions by contributing parts of their limited resources to different coalitions. Each coalition constitutes a group of players who are working together and whose utility is captured by both a coalition-level value and an individual user payoff. In a coalition formation game, each player individually decides which coalitions it wishes to join, so as to maximize its total payoff with the limited resources. Note that the coalitions can, in general, be overlapping, such that a player can participate in multiple coalitions simultaneously.

For the DCS problem, the players are the SUs $\mathcal{N} = \{1, 2, \ldots, N\}$ with power resource P_{SU} and bandwidth resource θ_{SU}. Here, a coalition $\mathcal{R}_i \subseteq \mathcal{N}$ denotes a cooperative group of SUs in which the coalition members report their sensing results to a given SU $i \in \mathcal{N}, i \in \mathcal{R}_i$. The power and bandwidth resources contributed by player $j \neq i, j \in \mathcal{R}_i$ to coalition $\mathcal{R}_i$ are $(\gamma_0 \sigma_u^2)/(\kappa d_{i,j}^{-\mu})$ and θ_0, respectively. Also, player $i \in \mathcal{N}$ naturally belongs to coalition $\mathcal{R}_i$ without contributing any power or bandwidth resource. Note that coalitions $\mathcal{R}_i$ and $\mathcal{R}_j$, $j \neq i$ can be exactly the same when SUs i and j receive the sensing results from the same SUs. However, we still treat them as two different coalitions and differentiate them with different subscripts, because:

(1) Coalitions $\mathcal{R}_i$ and $\mathcal{R}_j$ represent the received sensing results at different SUs, i.e., SU i and SU j.
(2) For any SU k belonging to both coalitions, coalitions $\mathcal{R}_i$ and $\mathcal{R}_j$ require different power resource contributions, i.e., $(\gamma_0 \sigma_u^2)/(\kappa d_{i,k}^{-\mu})$ and $(\gamma_0 \sigma_u^2)/(\kappa d_{j,k}^{-\mu})$.

For all N SUs in the network, there are exactly N coalitions that correspond to them. For the completeness of the game model, we define a *coalitional structure* as the set of all coalitions, denoted by $\mathcal{CS} = \{\mathcal{R}_1, \mathcal{R}_2, \ldots, \mathcal{R}_N\}$. Note that $\mathcal{CS}$ is just another expression of the reporting matrix $\mathbf{\Omega}$.

To capture the performance of a given coalition $\mathcal{R}_i$, we use a utility function that captures the best sensing performance of SU i, given by:

$$
U(\mathcal{R}_i) = \begin{cases} 2 - \min\limits_{\mathbf{\Lambda}(\mathcal{R}_i)} (Q_{m,i} + Q_{f,i}), & Q_m + Q_f, \\ 1 - \min\limits_{\mathbf{\Lambda}(\mathcal{R}_i), Q_{f,i} \leq \alpha} Q_{m,i}, & Q_m / Q_f, \end{cases} \tag{3.12}
$$

where $Q_{m,i}$ and $Q_{f,i}$ are given by (3.6) and (3.7) with $k_i = |\mathcal{R}_i|$, and $\mathbf{\Lambda}(\mathcal{R}_i)$ is the local sensing threshold vector for the players in $\mathcal{R}_i$. Note that we use "2−" and "1−" to maintain the utility to be positive, since all the probabilities $Q_{m,i}$ and $Q_{f,i}$ are between 0 and 1. Due to the symmetry of $U(\mathcal{R}_i)$ to the members in $\mathcal{R}_i$, we point out that the optimal value is obtained when all the coalition members have the same local sensing threshold, i.e., $\mathbf{\Lambda}(\mathcal{R}_i) = (\lambda, \lambda, \ldots, \lambda)_{1 \times |\mathcal{R}_i|}$.

For the $Q_m + Q_f$ criterion, by substituting (3.2), (3.3) and $k_i = |\mathcal{R}_i|$ into (3.6) and (3.7), and then substituting (3.6), (3.7) and $\mathbf{\Lambda}(\mathcal{R}_i) = (\lambda, \lambda, \ldots, \lambda)_{1 \times |\mathcal{R}_i|}$ into (3.12), the utility function of the $Q_m + Q_f$ criterion is given as:

$$
U(\mathcal{R}_i) = 1 - \min_\lambda \left\{ \left[\mathcal{Q} \left((\lambda - 1) \sqrt{N_s} \right) \right]^{|\mathcal{R}_i|} - \left[\mathcal{Q} \left(\left(\frac{\lambda}{1+\gamma} - 1 \right) \sqrt{N_s} \right) \right]^{|\mathcal{R}_i|} \right\}. \tag{3.13}
$$

The optimal λ_a is the zero point of the first first-order derivative, and thus, it satisfies:

$$
\frac{\partial}{\partial \lambda_a} \left[\mathcal{Q} \left(\left(\frac{\lambda_a}{1+\gamma} - 1 \right) \sqrt{N_s} \right) \right]^{|\mathcal{R}_i|} = \frac{\partial}{\partial \lambda_a} \left[\mathcal{Q} \left((\lambda_a - 1) \sqrt{N_s} \right) \right]^{|\mathcal{R}_i|}. \tag{3.14}
$$

By substituting $Q'(x) = \exp\left(-x^2/2\right)/\sqrt{2\pi}$, we have:

$$\left[\frac{Q\left((\lambda_a - 1)\sqrt{N_s}\right)}{Q\left(\left(\frac{\lambda_a}{1+\gamma} - 1\right)\sqrt{N_s}\right)}\right]^{|\mathcal{R}_i|-1} -$$

$$\frac{1}{1+\gamma}\exp\left\{\frac{N_s}{2}\left[(\lambda_a - 1)^2 - \left(\frac{\lambda_a}{1+\gamma} - 1\right)^2\right]\right\} = 0. \qquad (3.15)$$

Using (3.15), the optimal threshold of the $Q_m + Q_f$ criterion can be evaluated numerically, denoted by $\lambda_a(|\mathcal{R}_i|)$. Note that the solution is only decided by the coalition size $|\mathcal{R}_i|$.

By substituting $\lambda = \lambda_a(|\mathcal{R}_i|)$ into (3.13), the utility function in (3.12) for the $Q_m + Q_f$ criterion is also only determined by the coalition size $|\mathcal{R}_i|$, given by:

$$f_a(|\mathcal{R}_i|) = 1 - \left[Q\left((\lambda_a(|\mathcal{R}_i|) - 1)\sqrt{N_s}\right)\right]^{|\mathcal{R}_i|}$$

$$+ \left[Q\left(\left(\frac{\lambda_a(|\mathcal{R}_i|)}{1+\gamma} - 1\right)\sqrt{N_s}\right)\right]^{|\mathcal{R}_i|}. \qquad (3.16)$$

For the Q_m/Q_f criterion, by substituting (3.2), (3.3) and $k_i = |\mathcal{R}_i|$ into (3.6) and (3.7), and then substituting (3.6), (3.7) and $\Lambda(\mathcal{R}_i) = (\lambda, \lambda, \ldots, \lambda)_{1 \times |\mathcal{R}_i|}$ into (3.12), the utility function of the Q_m/Q_f criterion is given as:

$$U(\mathcal{R}_i) = \max_{\lambda}\left[Q\left(\left(\frac{\lambda}{1+\gamma} - 1\right)\sqrt{N_s}\right)\right]^{|\mathcal{R}_i|} \qquad (3.17a)$$

$$s.t. \left[Q\left((\lambda - 1)\sqrt{N_s}\right)\right]^{|\mathcal{R}_i|} \leq \alpha. \qquad (3.17b)$$

Note that $Q(x)$ is a decreasing function with its value between 0 and 1. We can solve the constraint inequality as:

$$\lambda \geq \lambda_{min} = 1 + \frac{Q^{-1}\left(\alpha^{1/|\mathcal{R}_i|}\right)}{\sqrt{N_s}}. \qquad (3.18)$$

Also, since $Q(x)$ is a decreasing function, the optimal threshold λ_b is the minimal value λ_{min}, given by:

$$\lambda_b(|\mathcal{R}_i|) = 1 + \frac{Q^{-1}\left(\alpha^{1/|\mathcal{R}_i|}\right)}{\sqrt{N_s}}. \qquad (3.19)$$

By substituting $\lambda = \lambda_b(|\mathcal{R}_i|)$ into the objective function, we have the utility function in (3.12) for the Q_m/Q_f criterion:

$$f_b(|\mathcal{R}_i|) = \left[\mathcal{Q}\left(\frac{1}{1+\gamma}\left(\mathcal{Q}^{-1}(\alpha^{1/|\mathcal{R}_i|}) - \gamma\sqrt{N_s} \right) \right) \right]^{|\mathcal{R}_i|}. \tag{3.20}$$

Given the optimal threshold $\lambda_a(|\mathcal{R}_i|)$ and the value of (3.12) $f_a(|\mathcal{R}_i|)$ for the $Q_m + Q_f$ criterion, as well as the optimal threshold $\lambda_b(|\mathcal{R}_i|)$ and the value of (3.12) $f_b(|\mathcal{R}_i|)$ for the Q_m/Q_f criterion, we have

$$\mathbf{\Lambda}(\mathcal{R}_i) = \begin{cases} (\lambda_a(|\mathcal{R}_i|), \dots, \lambda_a(|\mathcal{R}_i|))_{1 \times |\mathcal{R}_i|}, & Q_m + Q_f, \\ (\lambda_b(|\mathcal{R}_i|), \dots, \lambda_b(|\mathcal{R}_i|))_{1 \times |\mathcal{R}_i|}, & Q_m/Q_f, \end{cases} \tag{3.21}$$

and

$$U(\mathcal{R}_i) = U(|\mathcal{R}_i|) = \begin{cases} f_a(|\mathcal{R}_i|), & Q_m + Q_f, \\ f_b(|\mathcal{R}_i|), & Q_m/Q_f. \end{cases} \tag{3.22}$$

Note that $U(\mathcal{R}_i)$ is only determined by the coalition size, and its value is limited and discrete. The numerical results in Fig. 3.2 show that $U(|\mathcal{R}_i|)$ is an increasing concave function in both criteria, i.e.,

$$U(|\mathcal{R}_i|) > U(|\mathcal{R}_j|), \text{ with } |\mathcal{R}_i| > |\mathcal{R}_j|, \tag{3.23}$$

and

$$U(|\mathcal{R}_i|) - U(|\mathcal{R}_i| - 1) < U(|\mathcal{R}_j|) - U(|\mathcal{R}_j| - 1),$$
$$\text{with } |\mathcal{R}_i| > |\mathcal{R}_j|. \tag{3.24}$$

The utility function (3.12) captures the sensing performance of SU i when all members in $\mathcal{R}_i$ report to SU i by using the corresponding power and bandwidth resources. The network sensing performance, which is the average value of the SUs' sensing performance, therefore, is captured by the social welfare, defined as the sum utility of all the coalitions, given by:

$$\Upsilon(\mathcal{CS}) = \sum_{\mathcal{R}_i \in \mathcal{CS}} U(\mathcal{R}_i). \tag{3.25}$$

Considering the monotone-increasing property of $U(\cdot)$, as given by (3.23), we can expect a larger social welfare, or equally, a better network sensing performance, as the average coalition size increases. However, the power cost for a SU joining a coalition increases with the distance between the SU and the coalition. Thus, due to the limited power of each SU, the *grand coalition* that includes all SUs seldom forms.

For any player $j \neq i$, $j \in \mathcal{R}_i$, the payoff from coalition $\mathcal{R}_i$ is defined by the marginal utility due to player j's joining, given by:

$$\phi_j(\mathcal{R}_i) = U(\mathcal{R}_i) - U(\mathcal{R}_i \setminus \{j\}) = U(|\mathcal{R}_i|) - U(|\mathcal{R}_i| - 1), \tag{3.26}$$

Fig. 3.2 Coalition utility as
a function of coalition size
for both the $Q_m + Q_f$ and
Q_m/Q_f criteria. **a** The
$Q_m + Q_f$ criterion. **b** The
Q_m/Q_f criterion

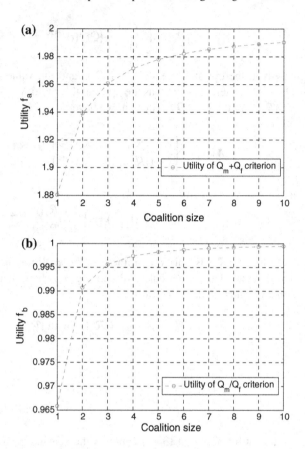

the payoff of player i is the remaining utility after coalition $\mathcal{R}_i$ pays all the other
members, i.e.,

$$\phi_i(\mathcal{R}_i) = U(|\mathcal{R}_i|) - (|\mathcal{R}_i| - 1)\,[U(|\mathcal{R}_i|) - U(|\mathcal{R}_i| - 1)]. \tag{3.27}$$

Due to the monotone-increasing property and the concavity of the utility function,
as given by (3.23) and (3.24), all the payoffs are positive, and only determined by
the coalition size $|\mathcal{R}_i|$. The numerical results in Fig. 3.3 show that $\phi_j(|\mathcal{R}_i|)$, $j \neq i$
is a decreasing convex function in both criteria, i.e.,

$$\phi_j(|\mathcal{R}_x|) < \phi_j(|\mathcal{R}_y|),$$
$$\text{with } |\mathcal{R}_x| > |\mathcal{R}_y|, \tag{3.28}$$

Fig. 3.3 Coalition payoff as a function of coalition size for both the $Q_m + Q_f$ and Q_m/Q_f criteria. **a** The $Q_m + Q_f$ criterion. **b** The Q_m/Q_f criterion

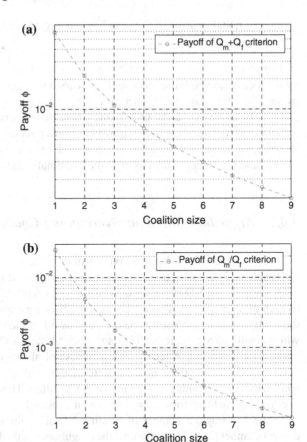

and

$$\phi_j(|\mathcal{R}_x| - 1) - \phi_j(|\mathcal{R}_x|) < \phi_j(|\mathcal{R}_y| - 1) - \phi_j(|\mathcal{R}_y|),$$
$$\text{with } |\mathcal{R}_x| > |\mathcal{R}_y|. \qquad (3.29)$$

For any given coalitional structure $\mathcal{CS}$, the total payoff of player $i \in \mathcal{N}$ is then given by:

$$\Phi_i(\mathcal{CS}) = \sum_{i \in \mathcal{R}_j, \mathcal{R}_j \in \mathcal{CS}} \phi_i(\mathcal{R}_j). \qquad (3.30)$$

Note that $\Phi_i(\mathcal{CS})$ is only determined by the sizes of the coalitions that player i participates in and the total payoff of all players is equal to the social welfare.

Definition 5 The *OCF game* is defined by the pair $(\mathcal{N}, U)$, where $\mathcal{N}$ is the set of players, and $U : 2^N \to \mathbb{R}$, given by (3.22), is the utility function. For any given coalitional structure $\mathcal{CS}$, the individual payoff of SU $i \in \mathcal{N}$ is $\Phi_i(\mathcal{CS})$, given by (3.30).

In the OCF-game model, the first step of the suboptimal solution is strictly captured by the local resource limitations, and the second steps (3.9) and (3.11) are captured by the utility function (3.12). Therefore, the centralized optimization problems (3.8) and (3.10) can be cast as the OCF game where the players choose their strategies in a distributed manner so as to maximize their own payoffs. As the individual payoffs increase, the social welfare also increases, and, in this case, the objective functions in (3.8) and (3.10) approach closer to their optimal values.

3.4.2 Algorithm Based on Overlapping Coalition Formation

We present a DCS algorithm that consists of three stages: (1) the neighbor discovery (ND) stage, (2) the coalition formation (CF) stage, and (3) the threshold decision (TD) stage. In the ND stage, each SU discovers nearby SUs as well as the distance to each of its neighbors. In the CF stage, the SUs communicate with each other via the control channel (reporting channel) and decide which SUs to report, or equally, which coalitions to join. In the TD stage, each SU decides its local sensing threshold using a local method. After the completeness of all the three stages, the SUs can perform DCS as described in the system model, with the reporting matrix and the local sensing thresholds given by the DCS algorithm. This DCS algorithm based on overlapping coalition formation is shown in Table 3.1.

In the ND stage, a number of existing ND algorithms can be applied over the control channel [36, 37]. We assume the neighbors within distance $\sqrt[\alpha]{(\kappa P_{SU})/(\gamma_0 \sigma_u^2)}$ are discovered, so that the received power at any SU is above κ_0 when its neighbor transmits at full power P_{SU}. The set of SU i's neighbors is denoted by $\mathcal{N}_i$. Note that the concept of neighbor is reciprocal. The distance $d_{i,j}$ between any two neighboring SU i and SU j is known by both ends.

In the CF stage, we give a coalition formation algorithm based on the OCF-game model. First, each SU initializes its state by joining as many coalitions as possible, i.e., each SU joins coalitions from the nearest to the farthest as long as its resource is sufficient. Formally, for SU $i \in \mathcal{N}$ with neighbors $n_1, n_2, \ldots, n_L, L = |\mathcal{N}_i|$, we assume $d_{i,n_j} \leq d_{i,n_{j+1}}, \forall 1 \leq j < L$. Then, SU i sequentially joins coalitions $\mathcal{R}_{n_1}, \mathcal{R}_{n_2}, \ldots, \mathcal{R}_{n_l}$ until the remaining power or bandwidth resource is insufficient for the next coalition $\mathcal{R}_{n_{l+1}}$, or it already joins all the nearby coalitions ($l = L$). Note that SU i naturally belongs to coalition $\mathcal{R}_i$ in all cases without contributing any power or bandwidth resource.

After the initialization, the SUs iteratively adjust their report-to sets $\mathcal{S}_i, i \in \mathcal{N}$ in a random order, so as to maximize their individual total payoff. Given the current coalitional structure $CS = \{\mathcal{R}_1, \mathcal{R}_2, \ldots, \mathcal{R}_N\}$, the best strategy of SU i is formulated as:

$$\max_{\mathcal{S}_i \subseteq \mathcal{N}_i} \sum_{j \in \mathcal{S}_i} \phi_i(\mathcal{R}_j \cup \{i\}), \tag{3.31a}$$

$$s.t. \sum_{j \in \mathcal{S}_i} \frac{\gamma_0 \sigma_u^2}{\kappa d_{i,j}^{-\mu}} \le P_{SU}, \tag{3.31b}$$

$$|\mathcal{S}_i|\theta_0 \le \theta_{SU}. \tag{3.31c}$$

Problem (3.31) is a knapsack problem with an extra constraint on the number of objects, which in most general cases is NP-complete [35]. Here, we use a "switch" operation for SUs to adjust their report-to sets, after which the total payoff of the considered SU is guaranteed to increase. The main idea of switch operation is to leave one low-paying coalition and join another high-paying coalition, as long as the remaining power can cover the possible extra consumption. The convergence of switch operations is proved in the next subsection.

Definition 6 For any given coalitional structure $CS = \{\mathcal{R}_1, \mathcal{R}_2, \ldots, \mathcal{R}_N\}$, a *switch* operation of player $i \in \mathcal{N}$ with remaining power P_i is defined by a pair $(\mathcal{R}_x, \mathcal{R}_y)$ that satisfies:

$$\frac{\gamma_0 \sigma_u^2}{\kappa d_{i,y}^{-\mu}} - \frac{\gamma_0 \sigma_u^2}{\kappa d_{i,x}^{-\mu}} \le P_i, \tag{3.32}$$

and

$$\phi_i(\mathcal{R}_y \cup \{i\}) > \phi_i(\mathcal{R}_x), \tag{3.33}$$

where $x, y \in \mathcal{N}_i$ and $i \in \mathcal{R}_x, i \notin \mathcal{R}_y$. For any SU $i \in \mathcal{N}$, a switch operation $(\mathcal{R}_x, \mathcal{R}_y)$ implies that SU i leaves coalition $\mathcal{R}_x$ and joins coalition $\mathcal{R}_y$.

In the TD stage, each coalition $\mathcal{R}_i$ seeks to find the optimal threshold vector $\Lambda(\mathcal{R}_i)$ in (3.21), so as to achieve the coalition utility as defined in (3.12). However, an SU may belong to multiple coalitions and the optimal threshold of one coalition is not necessarily the optimal threshold of the other coalitions. Therefore, we need a threshold decision algorithm for each SU to determine its practical sensing threshold. Generally speaking, this local threshold should be a function of the optimal thresholds $\lambda(\mathcal{R}_j)$ for all $\mathcal{R}_j$ including i. In the Q_m/Q_f criterion, in order to guarantee the false alarm probability, the SU should choose the maximum value of all the expected thresholds, i.e., $\max_{j|i \in \mathcal{R}_j} \lambda_b(|\mathcal{R}_j|)$. In the $Q_m + Q_f$ criterion, there are no constraints for false alarm or missed detection probabilities. Considering that each coalition represents the sensing performance of an SU, for fairness, the SU should choose the average value of all the expected thresholds, i.e., $[\sum_{j|i \in \mathcal{R}_j} \lambda_a(|\mathcal{R}_j|)][\sum_{j|i \in \mathcal{R}_j} 1]^{-1}$,

Table 3.1 DCS Algorithm based on Overlapping Coalition Formation

Neighbor Discovery Stage

For any SU $i \in \mathcal{N}$, it discovers its neighboring SUs within distance $\sqrt[\eta]{(\kappa P_{SU})/(\gamma_0 \sigma_u^2)}$, the set of which is denoted by $\mathcal{N}_i$, and the distance $d_{i,j}$ for any neighbor $j \in \mathcal{N}_i$.

Coalition Formation Stage

Each SU joins as many coalitions as possible by informing the corresponding SUs about its joning and the initial coalitional structure is given by $\mathcal{CS}_0$.

1: $\mathcal{CS} \leftarrow \mathcal{CS}_0$ % initial coalitional structure
2: **while** SU i has a switch operation $(\mathcal{R}_x, \mathcal{R}_y)$ as defined in Definition 6 **do**
3:     SU i informs SU x that it leaves coalition $\mathcal{R}_x$.
4:     SU x informs SUs $j \neq x$, $j \in \mathcal{R}_x \backslash \{i\}$ that SU i leaves coalition $\mathcal{R}_x$.
5:     The corresponding SUs update their information about coalition $\mathcal{R}_x \leftarrow \mathcal{R}_x \backslash \{i\}$.
6:     SU i informs SU y that it joins coalition $\mathcal{R}_y$.
7:     SU y informs SUs $j \neq y$, $j \in \mathcal{R}_y$ that SU i joins coalition $\mathcal{R}_y$.
8:     The corresponding SUs update their information about coalition $\mathcal{R}_y \leftarrow \mathcal{R}_y \cup \{i\}$.
9: **end while**
10: $\mathcal{CS}_f \leftarrow \mathcal{CS}$ % final coalitional structure

Threshold Decision Stage

For each SU $i \in \mathcal{N}$, the local sensing threshold λ_i is given by (3.34) with the current coalitional structure $\mathcal{CS}_f$.

where $[\sum_{j|i \in \mathcal{R}_j} 1]$ is the number of coalitions that SU i joins. Thus, for any final coalitional structure $\mathcal{CS}_f = \{\mathcal{R}_1, \mathcal{R}_2, \ldots, \mathcal{R}_N\}$, the local sensing threshold of SU $i \in \mathcal{N}$ is formally given by:

$$\lambda_i = \begin{cases} \left[\sum_{j|i \in \mathcal{R}_j} \lambda_a(|\mathcal{R}_j|) \right] \left[\sum_{j|i \in \mathcal{R}_j} 1 \right]^{-1}, & Q_m + Q_f \\ \max_{j|i \in \mathcal{R}_j} \lambda_b(|\mathcal{R}_j|), & Q_m/Q_f \end{cases} \tag{3.34}$$

where $\lambda_a(\cdot)$ and $\lambda_b(\cdot)$ are given in (3.15) and (3.19), respectively.

3.4.3 Convergence and Overhead

Theorem 1 *In the OCF game with any initial coalitional structure $\mathcal{CS}_0$, the network converges to a final coalitional structure $\mathcal{CS}_f$ within $\lceil E/\varepsilon \rceil$ switch operations, where $E = \sum_{i \in \mathcal{N}} U(|\mathcal{N}_i|) - \sum_{\mathcal{R}_i \in \mathcal{CS}_0} U(|\mathcal{R}_i|)$ and $\varepsilon = 2U(N-1) - U(N) - U(N-2)$.*

Proof For any current coalitional structure $\mathcal{CS}$, the utilities of coalitions $\mathcal{R}_x$ and $\mathcal{R}_y$ are changed after switch operation $(\mathcal{R}_x, \mathcal{R}_y)$ of SU i, while the utilities of the other coalitions remain the same. For coalition $\mathcal{R}_x$, its size decreases from $|\mathcal{R}_x|$ to

$|\mathcal{R}_x| - 1$ and its utility decreases from $U(|\mathcal{R}_x|)$ to $U(|\mathcal{R}_x| - 1)$. For coalition $\mathcal{R}_y$, its size increases from $|\mathcal{R}_y|$ to $|\mathcal{R}_y| + 1$ and its utility increases from $U(|\mathcal{R}_y|)$ to $U(|\mathcal{R}_y| + 1)$. Thus, the social welfare of the new coalitional structure $\mathcal{CS}'$ is given by:

$$
\begin{aligned}
\Upsilon(\mathcal{CS}') &= \Upsilon(\mathcal{CS}) - \left[U(|\mathcal{R}_x|) + U(|\mathcal{R}_y|)\right] \\
&\quad + \left[U(|\mathcal{R}_x| - 1) + U(|\mathcal{R}_y| + 1)\right] \\
&= \Upsilon(\mathcal{CS}) + \left[U(|\mathcal{R}_y| + 1) - U(|\mathcal{R}_y|)\right] \\
&\quad - [U(|\mathcal{R}_x|) - U(|\mathcal{R}_x| - 1)] \\
&= \Upsilon(\mathcal{CS}) + \phi_i(|\mathcal{R}_y \cup \{i\}|) - \phi_i(|\mathcal{R}_x|) \\
&> \Upsilon(\mathcal{CS})
\end{aligned}
\tag{3.35}
$$

Inequality (3.35) shows that a switch operation always increases the social welfare. Since the payoff function is a convex decreasing function, as given in (3.28) and (3.29), we have $\Upsilon(\mathcal{CS}') - \Upsilon(\mathcal{CS}) = \phi_i(|\mathcal{R}_y \cup \{i\}|) - \phi_i(|\mathcal{R}_x|) \geq \phi_i(N-1) - \phi_i(N) = 2U(N-1) - U(N) - U(N-2)$. Thus, we have a lower bound of the marginal increase of social welfare due to a single switch operation $\varepsilon = 2U(N-1) - U(N) - U(N-2)$. Also, the coalition utility is an increasing function, as given in (3.23), we have an upper bound of social welfare when each coalition $\mathcal{R}_i \subseteq \mathcal{N}_i$ reaches its largest size $|\mathcal{N}_i|$, given by $\sum_{i \in \mathcal{N}} U(|\mathcal{N}_i|)$. Thus, the gap of social welfare between $\mathcal{CS}_0$ and $\mathcal{CS}_f$ is limited by the upper bound $E = \sum_{i \in \mathcal{N}} U(|\mathcal{N}_i|) - \sum_{\mathcal{R}_i \in \mathcal{CS}_0} U(|\mathcal{R}_i|)$. Therefore, the network must converge within $\lceil E/\varepsilon \rceil$ switch operations. $\square$

Traditionally, the stability of OCSs is studied by the notion of *c-core*, in which an OCS is stable if no subset of players has the motivation to deviate from the current OCS and form new coalitions among themselves [30]. However, the notion *c-core* is based on the assumption that the deviators (players who remove their contribution from some of their coalitions) are untrustworthy and all coalitions should punish them by giving no payoff to them. In the OCF game, the players do not exhibit this property. In contrast, for our game, the deviators will not suffer any punishment. Thus, we need to define new notions to characterize the stability of the final OCS in the given algorithm.

Definition 7 In the OCF game, OCS $\mathcal{CS}$ is *switch-stable* if there does not exist a switch operation $(\mathcal{R}_x, \mathcal{R}_y)$ for any SU $i \in \mathcal{N}$ as defined in Definition 6.

For the algorithm given in Table 3.1, we directly have:

Lemma 1 *The final coalitional structure $\mathcal{CS}_f$ resulting from the algorithm in Table 3.1 is switch-stable.*

In general, the final coalitional structure $\mathcal{CS}_f$ is not the optimal solution. Also, the specific form of $\mathcal{CS}_f$ greatly depends on the sequence that the SUs perform switch operations and it is generally not unique. However, we still have the following proposition.

Proposition 2 *For any given CR network, let CS_{opt} denote the optimal coalitional structure with the highest social welfare $\Upsilon(CS_{opt})$, and let CS_0 and CS_f denote the initial and final coalitional structures in the overlapping algorithm. We have:*

$$\frac{\Upsilon(CS_f)}{\Upsilon(CS_{opt})} \geq \frac{\sum_{\mathcal{R}_i \in CS_0} U(|\mathcal{R}_i|)}{NU(\lceil \sum_{\mathcal{R}_i \in CS_0} |\mathcal{R}_i|/N \rceil)}. \tag{3.36}$$

Proof Since the utility function $U(\cdot)$ is an increasing concave function, as given in (3.23) and (3.24), the optimal social welfare satisfies:

$$\Upsilon(CS_{opt}) = \sum_{\mathcal{R}_i \in CS_{opt}} U(|\mathcal{R}_i|) \leq NU(\lceil \sum_{\mathcal{R}_i \in CS_{opt}} |\mathcal{R}_i|/N \rceil). \tag{3.37}$$

In the overlapping algorithm, as we noted, each SU joins as many coalitions as possible in the initialization period of the CF stage. Therefore, the initial coalitional structure CS_0 has the largest sum coalition size among all the feasible coalitional structures. Note that a switch operation does not change the sum size of the involved coalitions. We have:

$$\sum_{\mathcal{R}_i \in CS_f} |\mathcal{R}_i| = \sum_{\mathcal{R}_i \in CS_0} |\mathcal{R}_i| \geq \sum_{\mathcal{R}_i \in CS_{opt}} |\mathcal{R}_i|. \tag{3.38}$$

Since $U(\cdot)$ is an increasing function, as given in (3.23), by substituting (3.38) into (3.37), we have:

$$\Upsilon(CS_{opt}) \leq NU(\lceil \sum_{\mathcal{R}_i \in CS_0} |\mathcal{R}_i|/N \rceil). \tag{3.39}$$

Note that after a switch operation, the social welfare strictly increases. We have:

$$\Upsilon(CS_f) \geq \Upsilon(CS_0) = \sum_{\mathcal{R}_i \in CS_0} U(|\mathcal{R}_i|). \tag{3.40}$$

Combining (3.39) and (3.40), we have:

$$\frac{\Upsilon(CS_f)}{\Upsilon(CS_{opt})} \geq \frac{\sum_{\mathcal{R}_i \in CS_0} U(|\mathcal{R}_i|)}{NU(\lceil \sum_{\mathcal{R}_i \in CS_0} |\mathcal{R}_i|/N \rceil)}. \tag{3.41}$$

$\square$

Proposition 2 shows that, in the given algorithm, the relative performance of the final coalitional structure CS_f, compared with the optimal coalitional structure, is guaranteed to be above a certain threshold. This threshold only depends on the initial coalitional structure CS_0 given by the initialization process in the coalition formation stage. For a given CR network, the initialization process generates a unique coalitional structure CS_0. Thus, the threshold is only determined by the network parameters, and therefore, the relative performance, compared with the optimal solution, is guaranteed.

The overhead required for practically implementing the algorithm in Table I mainly relates to the stage in which the SUs initialize their states as well as when a switch operation is performed. We assume an SU's identity can be represented by τ bits. Note that each coalition corresponds to a particular SU. A coalition's identity also requires τ bits. For the message that SU i leaves or joins coalition $\mathcal{R}_j$, by ignoring the 1 bit to distinguish "leave" and "join," we can transmit this message in a packet of 2τ bits.

In the initialization of coalitional structure $\mathcal{CS}_0$, each SU $i \in \mathcal{N}$ receives the information from SU $j \neq i$, $j \in \mathcal{R}_i$ that SU j joins coalition $\mathcal{R}_i$. Thus, the overhead for initialization is given by:

$$T_{init}(\mathcal{CS}_0) = \sum_{\mathcal{R}_i \in \mathcal{CS}_0} 2(|\mathcal{R}_i| - 1)\tau. \tag{3.42}$$

For performing a switch operation $(\mathcal{R}_x, \mathcal{R}_y)$, SU i informs SU x that SU i it wishes to leave coalition $\mathcal{R}_x$, and informs SU y that it will join coalition $\mathcal{R}_y$. Then, SU x and SU y update their coalition information by informing their coalition members about SU i's joining or leaving coalition $\mathcal{R}_x$ or $\mathcal{R}_y$. Thus, the overhead of switch operation $(\mathcal{R}_x, \mathcal{R}_y)$ is given by:

$$\begin{aligned} T_{switch}(\mathcal{R}_x, \mathcal{R}_y) &= 4\tau + 2\tau(|\mathcal{R}_x| - 2) + 2\tau(|\mathcal{R}_y| - 1) \\ &= 2\tau(|\mathcal{R}_x| + |\mathcal{R}_y| - 1). \end{aligned} \tag{3.43}$$

The coalition size is approximately $\mathcal{O}(N)$. Thus, the overhead of the initialization period is $\mathcal{O}(N^2)$, and the overhead of a single switch operation is $\mathcal{O}(N)$. Note that the network converges within $\lceil E/\varepsilon \rceil$ switch operations, as given by Theorem 1. The worst-case overhead is approximately $\mathcal{O}(N^2) + \lceil E/\varepsilon \rceil \mathcal{O}(N)$.

3.5 DCS Based on Nonoverlapping Coalition Formation Games

In this section, we extend the popular nonoverlapping CF-game model for cooperative sensing that is proposed in [27] while considering the power and bandwidth constraints and allowing the utility to reflect the Q_m/Q_f criterion as well as the $Q_m + Q_f$ criterion. Here, we reconsider the CF-game model with the newly defined coalition utility, and then, we point out its limitations when compared to the more general OCF-game model of Section IV.

3.5.1 Nonoverlapping CF-Game Model

In the nonoverlapping CF game, the players are also the SUs $\mathcal{N} = \{1, 2, \ldots, N\}$ with power resource P_{SU} and bandwidth resource θ_{SU}. The players form disjoint nonoverlapping coalitions and the coalitional structure $\mathcal{CS}$ is a *partition* of $\mathcal{N}$. Each player $i \in \mathcal{N}$ that belongs to coalition $\mathcal{C} \subseteq \mathcal{N}$ reports to the players in the same coalition by contributing power $\sum_{j \neq i, j \in \mathcal{C}} (\kappa_0 \sigma_u^2)/(A d_{i,j}^{-\mu})$ and bandwidth $(|\mathcal{C}| - 1)\theta_0$. Thus, each SU in $\mathcal{C}$ can receive the sensing data of all SUs in $\mathcal{C}$, and the utility of coalition $\mathcal{C}$ is thus:

$$V(\mathcal{C}) = \sum_{i \in \mathcal{C}} U(\mathcal{C}) = |\mathcal{C}|\, U(|\mathcal{C}|), \tag{3.44}$$

where $U(|\mathcal{C}|)$ is given by (3.22). Unlike the OCF game, the coalition in nonoverlapping CF game represents the sum performance of all its coalition members. To achieve the utility defined in (3.44), we have the optimal threshold vector given by:

$$\mathbf{\Lambda}(\mathcal{C}) = \begin{cases} (\lambda_a(|\mathcal{C}|), \ldots, \lambda_a(|\mathcal{C}|))_{1 \times |\mathcal{C}|}, & Q_m + Q_f \\ (\lambda_b(|\mathcal{C}|), \ldots, \lambda_b(|\mathcal{C}|))_{1 \times |\mathcal{C}|}, & Q_m/Q_f \end{cases} \tag{3.45}$$

where $\lambda_a(\cdot)$ and $\lambda_b(\cdot)$ are given in (3.15) and (3.19), respectively.

The social welfare is also defined as the sum utility of all the coalitions, given by

$$\Xi(\mathcal{CS}) = \sum_{\mathcal{C} \in \mathcal{CS}} |\mathcal{C}| U(|\mathcal{C}|) = \sum_{j \in \mathcal{N}|j \in \mathcal{C}} U(|\mathcal{C}|). \tag{3.46}$$

Since $U(|\mathcal{C}|)$ reflects the sensing performance of each SU in $\mathcal{C}$, then, the defined social welfare also reflects the network sensing performance. As similar as the OCF-game model, due to the monotone-increasing property of $U(\cdot)$, the network performs better as the average coalition size increases. Also, due to the increasing power cost for joining a larger coalition, the grand coalition may not always form.

We assume that the utility of each coalition is equally distributed to each coalition member, and the individual payoff of any player $i \in \mathcal{N}$ is then given by:

$$\Psi_i(\mathcal{CS}) = \psi_i(\mathcal{C}) = U(|\mathcal{C}|), \tag{3.47}$$

where $i \in \mathcal{C}$ and $\mathcal{C} \in \mathcal{CS}$. Note that the coalitions are completely disjoint and each SU belongs to only one coalition. The total payoff of an SU is the payoff from the coalition it belongs to. Naturally, the total payoff of all SUs is equal to the defined social welfare.

Definition 8 The *CF game* is defined by the pair $(\mathcal{N}, V)$, where $\mathcal{N}$ is the set of players, and $V : 2^N \to \mathbb{R}$, given by (3.44), is the utility function. For any given coalitional structure $\mathcal{CS}$, the individual payoff of SU $i \in \mathcal{N}$ is $\Psi_i(\mathcal{CS})$, given by (3.47).

Compared with the OCF-game model defined in Definition 5, the nonoverlapping CF-game model also captures the suboptimal solution by the local resource limitations and its newly defined utility function (3.44). Moreover, the optimal sensing threshold given by (3.45) is more practical since the coalitions are disjoint and each SU belongs to only one coalition. Therefore, the increase of individual payoff, or equally the increase of social welfare, means an equal increase of the objective functions in (3.8) and (3.10).

However, the nonoverlapping CF-game model imposes extra limitations on the reporting structure due to the nonoverlapping assumption. From the perspective of an OCF game, any coalition $\mathcal{C}$ in the nonoverlapping CF-game model represents $|\mathcal{C}|$ identical coalitions in the OCF-game model $\mathcal{R}_i, i \in \mathcal{C}$. Thus, the nonoverlapping CF-game model can be seen as a special case of the OCF-game model, in which the N overlapping coalitions are classified into groups, and in each group, the coalitions are identical to a coalition consisting of the SUs that these coalitions correspond to. Next, we show the limitation of the CF-game model via a special case. In Fig. 3.1, there are three nearby SUs $\{1, 2, 3\}$ and we assume each SU can only report to one SU due to the power and bandwidth constraints. In the OCF-game model, we can expect the coalitional structure $\mathcal{R}_1 = \{1, 3\}, \mathcal{R}_2 = \{2, 1\}, \mathcal{R}_3 = \{3, 2\}$ to form. Thus, the sensing performance of the SUs are respectively given by $U(2), U(2)$, and $U(2)$. In the nonoverlapping CF-game model, the network forms a structure with a two-SU coalition and a singleton, and, thus, the sensing performance will be given by $U(1), U(2)$ and $U(2)$. Clearly, the result of the OCF-game model strictly outperforms the nonoverlapping CF-game model.

3.5.2 Algorithm Based on Nonoverlapping Coalition Formation

In CF games, the merge-and-split algorithm is often used to achieve a stable coalitional structure [26, 27]. In this algorithm, multiple coalitions merge into one larger coalition and a single coalition split into multiple smaller coalitions, as long as the payoffs of all the involved players are increased. In the considered CF game, each player's payoff increases with the coalition size, as seen in (3.47). Thus, the players always prefer larger coalitions and the merge-and-split algorithm degrades to the merge algorithm where the coalitions keep merging until the bandwidth or power resource is completely used for some players. The DCS algorithm based on nonoverlapping coalition formation is formally given in Table 3.2.

In the ND stage, we use the same method as in the overlapping case, where the SUs within distance $\sqrt[\nu]{(\kappa P_{SU})/(\gamma_0 \sigma_u^2)}$ are discovered as the neighbors, the neighbor set of SU $i \in \mathcal{N}$ is also denoted by $\mathcal{N}_i$.

In the CF stage, we define the merge operation as follows:

Definition 9 Given the coalitional structure $\mathcal{CS}$, a *merge* operation in $\mathcal{CS}$ is defined by a pair $(\mathcal{C}_1, \mathcal{C}_2)$ of two disjoint coalitions that satisfies:

$$\sum_{j \neq i, j \in \mathcal{C}_1 \cup \mathcal{C}_2} \frac{\gamma_0 \sigma_u^2}{\kappa d_{i,j}^{-\mu}} \leq P_{SU}, i \in \mathcal{C}_1 \cup \mathcal{C}_2, \tag{3.48}$$

$$(|\mathcal{C}_1 \cup \mathcal{C}_2| - 1)\theta_0 \leq \theta_{SU}, \tag{3.49}$$

where any two SUs in $\mathcal{C}_1$ and $\mathcal{C}_2$ are neighbors, i.e., $i \in \mathcal{N}_j, \forall i, j \in \mathcal{C}_1 \cup \mathcal{C}_2$.

Suppose each coalition $\mathcal{C} \subset \mathcal{N}$ has a coalition head that has the complete information of all the coalition members, i.e., $\mathcal{N}_i, i \in \mathcal{C}$ and $d_{i,j}, i \in \mathcal{C}, j \in \mathcal{N}_i$. Therefore, the merge operation between two coalitions is actually performed by the two coalitions heads that represent them. Note that any feasible merge operation require all the involved players to be neighbors. Any coalition member can be chosen as the coalition head without missing any feasible merge operations. For the coalition formed by a merge operation, the coalition head is randomly chosen from the two original coalitions heads. For practical reasons, each coalition head maintains a tag parameter for each of its neighboring coalition heads. Formally, for any two neighboring heads i and j of coalitions $\mathcal{C}_1$ and $\mathcal{C}_2$, tags $t_{i,j} = 0$ and $t_{j,i} = 0$ represents that a merge operation $(\mathcal{C}_1, \mathcal{C}_2)$ is not feasible. A coalition head only tries the merge operations with nonzero tags. If an SU is no longer a coalition head, the corresponding tags are deleted.

In the TD stage, the optimal threshold vector $\Lambda(\mathcal{C})$ for any coalition $\mathcal{C} \in \mathcal{CS}_f$ is given by (3.45), where $\mathcal{CS}_f$ is the final coalition structure given by the CF stage. Thus, the local sensing threshold of SU $i \in \mathcal{C}, \mathcal{C} \in \mathcal{CS}_f$ is formally given by:

$$\lambda_i = \begin{cases} \lambda_a(|\mathcal{C}|), & \mathcal{Q}_m + \mathcal{Q}_f \\ \lambda_b(|\mathcal{C}|). & \mathcal{Q}_m / \mathcal{Q}_f \end{cases} \tag{3.50}$$

where $\lambda_a(\cdot)$ and $\lambda_b(\cdot)$ are given in (3.15) and (3.19), respectively.

3.5.3 Convergence and Overhead

The convergence of the nonoverlapping coalition formation algorithm is a direct result of the defined merge rule, and follows directly from known results such as [13, 26, 27]. Actually, we can expect the algorithm to converge within N merge operations, since each merge operation will decrease the number of coalitions by 1. In the algorithm in Table 3.2, the main source of overhead pertains to the case when a coalition tries to merge with another coalition and transmits the complete coalition information. If the merge operation is feasible and actually executed, the new coalition head updates coalition information with additional overhead, and the "retired" coalition head informs its original members about its "retirement" with 1 bit information. If the merge operation is not feasible, only the 1 bit fail information is transmitted. Here, we also assume an SU's identity requires τ bits and ignore the

Table 3.2 DCS Algorithm based on Nonoverlapping Coalition Formation

Neighbor Discovery Stage

For each SU $i \in \mathcal{N}$, it discovers its neighboring SUs within distance $\sqrt[\mu]{(\kappa P_{SU})/(\gamma_0 \sigma_u^2)}$, the set of which is denoted by $\mathcal{N}_i$, and the distance $d_{i,j}$ for any neighbor $j \in \mathcal{N}_i$.

Coalition Formation Stage

1: $\mathcal{CS} \leftarrow \{\{1\}, \{2\}, \ldots, \{N\}\}$ % each SU forms a singleton
2: $t_{i,j} \leftarrow 1, i \in \mathcal{N}, j \in \mathcal{N}_i$ % each SU maintains a tag corresponding to a neighbor coalition head
3: **while** SU $i \in C$ has a tag $t_{i,j} = 1$ **do**
4:    SU i sends SU j the complete information of coalition C.
5:    SU $j \in C'$ computes if (C, C') is a merge operation as defined in Definition 9.
6:    **if** (C, C') is a merge operation **then**
7:       % C and C' merge into $C \cup C'$ with coalition head j.
8:       SU j informs SUs $k \in C$ that SUs in C' join their coalition C, and sets tags $t_{j,k} \leftarrow 0$.
9:       SUs $k \in C$ update their coalition information and set all their tags to zero.
10:       SU j informs SUs $k \neq j, k \in C'$ that SUs in C join their coalition C', and sets tags $t_{j,k} \leftarrow 0$.
11:       SUs $k \neq j, k \in C'$ update their coalition information and set all their tags to zero.
12:    **else**
13:       SU j informs SU i that the trying (C, C') fails, and sets tag $t_{j,i} \leftarrow 0$.
14:       SU i sets its tag $t_{i,j} \leftarrow 0$.
15:    **end if**
16: **end while**
17: $\mathcal{CS}_f \leftarrow \mathcal{CS}$ % final coalitional structure

Threshold Decision Stage

For any SU $i \in \mathcal{N}$, the local sensing threshold λ_i is determined by (3.50) with the current coalitional structure $\mathcal{CS}_f$.

1 bit information. Note that in the CF-game model, each SU belongs to one and only one coalition. The information that SU i joins coalition C received by SU $j \in C$ can be represented by the identity of SU i without causing any ambiguity.

The complete information of coalition C includes the information of each coalition member $i \in C$, which consists of SU i itself, all its neighbors in $\mathcal{N}_i$ and the corresponding distances $d_{i,j}, j \in \mathcal{N}_i$. We simply assume a distance requires τ bits. Thus, the complete information of coalition C is given by:

$$T_{try}(C) = \sum_{i \in C} (2|\mathcal{N}_i| + 1). \tag{3.51}$$

In merge operation (C, C'), coalition head $j \in C'$ becomes the head of the merged coalition $C \cup C'$, and then, it informs the SUs in C about the joining of the SUs in C', as well as the SUs in C' (except for itself) about the joining of the SUs in C. Thus, the overhead of merge operation (C, C') is given by:

$$T_{merge}(\mathcal{C}, \mathcal{C}') = |\mathcal{C}| \times |\mathcal{C}'| + (|\mathcal{C}'| - 1) \times |\mathcal{C}|$$
$$= (2|\mathcal{C}'| - 1)|\mathcal{C}|\tau. \tag{3.52}$$

where SU i and SU j are coalition heads of coalitions $\mathcal{C}$ and $\mathcal{C}'$, respectively.

$T_{try}(\mathcal{C})$ is $\mathcal{O}(N^2)$, and $T_{merge}(\mathcal{C}, \mathcal{C}')$ is $\mathcal{O}(N^2)$. In Table 3.2, in each attempt for a merge operation, at least one tag parameter is set to zero. Thus, the network converges within N^2 attempts, and the total overhead of trying is less than $\mathcal{O}(N^4)$. Also, the network converges within N merge operations. Thus, the overhead of merge operations is $\mathcal{O}(N^3)$. Therefore, the total overhead is less than $\mathcal{O}(N^4)$.

3.6 Practical Issues

In the considered DCS problem, in order to simplify our discussion and calculation, some of the practical issues are not considered. In this section, we discuss how these practical issues may effect our proposal and how we can extend our model to involve these factors.

The utility function is the fundamental characterization of an OCF game. In the considered OCF game, the utility function $U(\mathcal{R}_i)$ precisely represents the sensing performance of SU i in both the $Q_m + Q_f$ and Q_m / Q_f criteria, as seen in (3.22). Due to the increasing monotony and concavity of $U(\mathcal{R}_i)$, the social welfare is guaranteed to be increased by each switch operation, and thus, the system sensing performance is always increasing until the network converges to a switch-stable outcome. In fact, as long as $U(\mathcal{R}_i)$ is defined as the best sensing performance we can achieve from this coalition, we can easily understand that any extra data only increases the coalition utility and the marginal improvement only decreases with the coalition size, i.e., $U(\mathcal{R}_i)$ is monotone increasing and concave.

In the system model, we assume that the PU is far away from the SUs, so that the received SNRs are the same for all SUs. However, we can replace the common received SNR γ with γ_i for each SU i, so as to consider small-scale scenarios in which the SUs have different distances to the PU. As we noted, $U(\mathcal{R}_i)$ is still monotone increasing and concave, and thus, the OCF algorithm can still work effectively, only that the calculation becomes more complex. For similar reasons, we can also extend our model to involve more practical concerns, such as more sophisticated fusion rules as in consensus-based algorithms [4], the spatial correlation between SUs [22, 23], the mobility of PUs [24], the nonidealness of the report channel [25], and even the location and time diversities of SUs in spectrum-heterogeneous systems [18–21]. For each practical issue, the utility function $U(\mathcal{R}_i)$ should be redefined to reflect such concern, but still should present the best sensing performance we can achieve for SU i. Note that these practical issues may complicate the expression of $U(\mathcal{R}_i)$, and we may need to simplify $U(\mathcal{R}_i)$ to reduce the computational complexity. However, no matter how we define $U(\mathcal{R}_i)$, the properties of increasing monotony and concavity should always be guaranteed.

3.7 Simulation Results and Analysis

For our simulations, we consider a network in which the SUs are randomly distributed within a $10\,\text{km} \times 10\,\text{km}$ square area and the PU is $D = 150\,\text{km}$ away from the square center. The path loss parameters are $\kappa = 1$ and $\mu = 3$, and the noise power is $\sigma_u^2 = -90\,\text{dBm}$. The PU transmit power σ_s^2 is set in such a way that the average received SNR at the SUs is $\gamma = -15\,\text{dB}$, and the number of samples at each SU is set to $N_s = 10{,}000$. For the power and bandwidth constraints, the minimum received SNR and minimum time-frequency resource for transmitting 1 bit are set to $\gamma_0 = 0\,\text{dB}$ and $\theta_0 = 1$, respectively. In the Q_m/Q_f criterion, the maximum false alarm constraint is set to $\alpha = 0.1$, as recommended by the IEEE 802.22 standard [38]. The remaining parameters are varied within given ranges so as to evaluate the performance of different algorithms under different conditions. All statistical results are averaged over the random locations of the SUs via a large number of independent runs.

3.7.1 Comparison of DCS Algorithms

In Fig. 3.4, we show the snapshots of coalitional structures of both the nonoverlapping and overlapping algorithms in a 5-SU network. The power and bandwidth constraints are set to $P_{SU} = 100\,\text{mW}$ and $\theta_{SU} = 10$. As we see, in the nonoverlapping algorithm, the SUs form a 3-coalition nonoverlapping structure $\mathcal{C}_1 = \{2\}, \mathcal{C}_2 = \{1, 4\}, \mathcal{C}_3 = \{3, 5\}$, and each SU reports to the SUs in the same coalition. While, in the overlapping algorithm, the SUs form a 5-coalition overlapping structure $\mathcal{R}_1 = \{1, 2\}, \mathcal{R}_2 = \{1, 2, 4\}, \mathcal{R}_3 = \{3, 5\}, \mathcal{R}_4 = \{2, 4\}, \mathcal{R}_5 = \{3, 5\}$, and in each coalition, all the members report to the particular SU that the coalition corresponds to.

In Fig. 3.5, we show the network sensing performance as a function of the network size N in both the $Q_m + Q_f$ and Q_m/Q_f criteria. The power and bandwidth constraints are set as $P_{SU} = 100\,\text{mW}$ and $\theta_{SU} = 10$. It shows that, for both criteria, the cooperative algorithms outperform the local spectrum sensing, and their cooperative gains increase with the network size. Also, the overlapping DCS outperforms the nonoverlapping DCS in all cases, and the gap between them increases with the network size. When the network is sparse, both cooperative algorithms have similar performance. While, when the network is dense ($N = 50$), the total error probability ($Q_m + Q_f$ criterion) is reduced from 0.04 to 0.01, which is 25 %, and the missed detection probability (Q_m/Q_f criterion) is reduced from 0.005 to 0.001, which is 20 %. As the network becomes denser, each SU can cooperate with more neighbors with the same power and bandwidth resources, and thus, the average coalition size increases. In both Sections IV and V, the network sensing performance is represented by the social welfare, which increases with the average coalition size. Therefore, the increasing network size can improve the network sensing performance by increasing the average coalition size, as seen in Fig. 3.5.

Fig. 3.4 Snapshots of the coalitional structures resulting from both the nonoverlapping and overlapping algorithms in a 5-SU network with power constraint $P_{SU} = 100\,\text{mW}$ and bandwidth constraint $\theta_{SU} = 10$. **a** Non-overlapping algorithm. **b** Overlapping algorithm

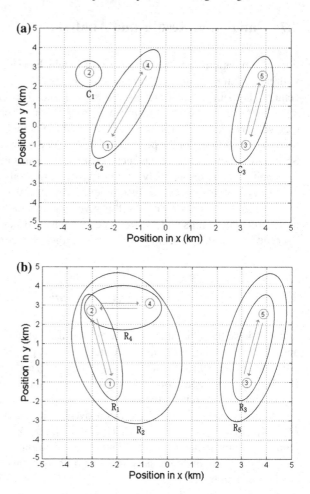

In Fig. 3.6, we show the average coalition size as a function of network size N resulting from both cooperative algorithms. It clearly shows that the overlapping DCS achieves a much larger coalition size than the nonoverlapping DCS, which explains the performance gap as seen in Fig. 3.5. In general, the overlapping structure provides the SUs with more flexibility on the distribution of their local resources, which encourages them to cooperate with more neighbors, and thus, increases the average coalitions size and improves the network sensing performance. Figure 3.6 shows that the average coalition size for the overlapping case reaches a maximum of 11 for a network with $N = 50$ SUs, while that for the nonoverlapping case does not exceed 4.

In Fig. 3.7, we show the probability density functions of coalition size per SU for both cooperative algorithms. It shows that in the nonoverlapping algorithm, the coalitions with sizes 2, 3, 4 occupy about 80 % of all coalitions, while in the over-

Fig. 3.5 Network sensing
performance as a function of
the number of SUs N with
power constraint
$P_{SU} = 100\,\text{mW}$ and
bandwidth constraint
$\theta_{SU} = 10$ in both criteria.
The false alarm constraint in
the Q_m/Q_f criterion is
$\alpha = 0.1$. **a** The $Q_m + Q_f$
criterion. **b** The Q_m/Q_f
criterion

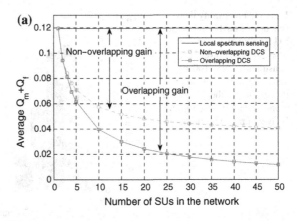

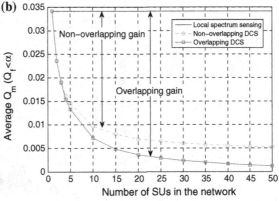

Fig. 3.6 Average coalition
size as a function of the
number of SUs N with power
constraint $P_{SU} = 100\,\text{mW}$
and bandwidth constraint
$\theta_{SU} = 10$ in either criteria

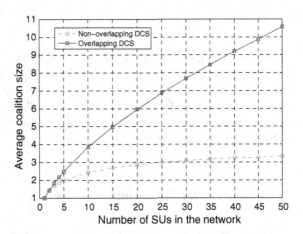

Fig. 3.7 Probability density
function of coalition size per
SU for networks with
$N = 30$ SUs, power
constraint $P_{SU} = 100\,\text{mW}$
and bandwidth constraint
$\theta_{SU} = 10$ in either criteria. **a**
Non-overlapping algorithm.
b Overlapping algorithm

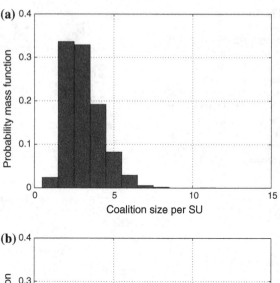

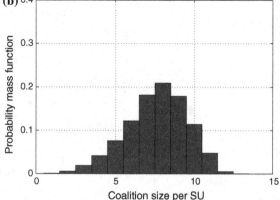

lapping algorithm, coalitions with sizes 6, 7, 8, 9, 10 occupy the same percentage.
This is in line with the result in Fig. 3.6 which shows that the overlapping algorithm
forms larger coalitions. In addition, this also implies that the variance of the sizes of
the coalitions resulting from the overlapping algorithm exceeds those resulting from
the nonoverlapping algorithm. Thus, the SUs in the overlapping algorithm have a
wider range of sensing performance.

3.7.2 Power and Bandwidth Constraints

Figures 3.8 and 3.9 show the network sensing performance as a function of the bandwidth θ_{SU} and the power P_{SU}, respectively. In Fig. 3.8, we show the curves of infinite
power ($P_{SU} = \infty$) and limited power ($P_{SU} = 60\,\text{mW}$) for each algorithm in each
criterion. The gap between the two curves represents the corresponding performance

Fig. 3.8 Network sensing performance as a function of the total bandwidth of each SU θ_{SU} with network size $N = 30$. The false alarm constraint in the Q_m/Q_f criterion is $\alpha = 0.1$. **a** The $Q_m + Q_f$ criterion. **b** The Q_m/Q_f criterion

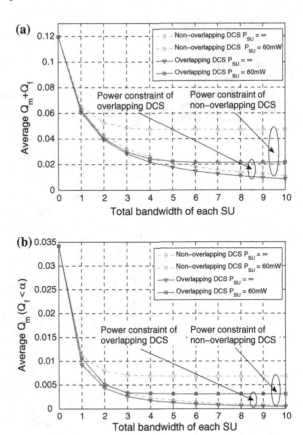

decrease due to the power constraint. As similar to Fig. 3.8, we show performance decrease due to the bandwidth constraint in Fig. 3.9, by showing the curves of infinite bandwidth ($\theta_{SU} = \infty$) and limited bandwidth ($\theta_{SU} = 2$). Figure 3.8 (Fig. 3.9) clearly shows that the curves of infinite power (bandwidth) decrease as the bandwidth (power) resource increases, while the curves of limited power (bandwidth) flatten out once the bandwidth (power) resource exceeds a certain threshold. Thus, the gap due to the power (bandwidth) constraint increases with the bandwidth (power) resource. In the $Q_m + Q_f$ criterion, when the bandwidth (power) is sufficient $\theta_{SU} = 10$ ($P_{SU} = 100\,\text{mW}$), the power (bandwidth) constraint increases the total error probability from 0.01 (0.044) to 0.05 (0.052) of the nonoverlapping algorithm, and from 0.008 (0.02) to 0.02 (0.04) of the overlapping algorithm. In the Q_m/Q_f criterion, when the bandwidth (power) is sufficient $\theta_{SU} = 10$ ($P_{SU} = 100\,\text{mW}$), the power (bandwidth) constraint increases the missed detection probability from 0.001 (0.006) to 0.007 (0.008) of the nonoverlapping algorithm, and from 0.001 (0.002) to 0.003 (0.005) of the overlapping algorithm.

Fig. 3.9 Network sensing performance as a function of the total power of each SU P_{SU} with network size $N = 30$. The false alarm constraint in the Q_m/Q_f criterion is $\alpha = 0.1$. **a** The $Q_m + Q_f$ criterion. **b** The Q_m/Q_f criterion

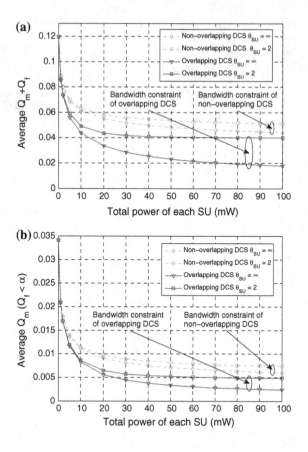

The behavior of the curves shown in Figs. 3.8 and 3.9 can be explained as follows. As previously noted, the network sensing performance is mainly determined by the average coalition size, which in general is limited by both power and bandwidth constraints. If the power (bandwidth) is infinite, the bandwidth (power) becomes the only limitation and the performance monotonously improves with the increasing bandwidth (power), as seen in the curves of infinite power (bandwidth) in Fig. 3.8 (Fig. 3.9). If the power (bandwidth) is limited, it will become the major limitation when the bandwidth (power) is sufficiently large, and even if we keep increasing the bandwidth (power) resource, the performance keeps stationary, as we see in the curves of limited power (bandwidth) in Fig. 3.8 (Fig. 3.9). Therefore, the gap due to the power (bandwidth) constraint increases with the bandwidth (power) resource, as seen in Fig. 3.8 (Fig. 3.9).

In Fig. 3.10, we show the average resource utilization as a function of the number of SUs N for both bandwidth and power resources. When the network is dense ($N = 50$), 90 % power and 90 % bandwidth are utilized by the overlapping algorithm, while only 50 % power and 25 % bandwidth are utilized by the nonoverlapping algorithm. As we noted, the overlapping structure provides the SUs with more flexibility on the

Fig. 3.10 Average resource utilization as a function of the number of SUs N with power constraint $P_{SU} = 100$ mW and bandwidth constraint $\theta_{SU} = 10$ in either criteria. The false alarm constraint in the Q_m / Q_f criterion is $\alpha = 0.1$. **a** Power utilization. **b** Bandwidth utilization

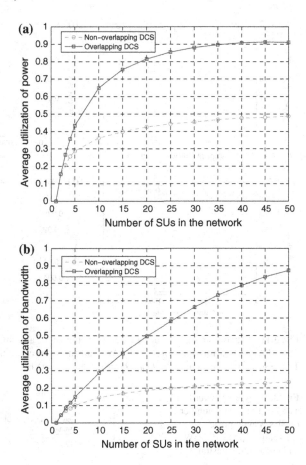

distribution of local resources, which enables them to contribute more resources, and thus, increases the power and bandwidth utilizations, as seen in Fig. 3.10. The higher resource utilization of the overlapping algorithm increases the average coalition size, and thus, improves the network sensing performance, as seen in Figs. 3.5 and 3.6.

3.7.3 Convergence, Overhead, and Complexity

In Fig. 3.11, we show the network sensing performance as a function of the maximum overhead for each algorithm in each criterion for networks with $N = 30$ SUs. For both the cooperative algorithms, we can see that the network sensing performance improves fast as the SUs begin to exchange information, and then converges steadily to a final value. For the overlapping algorithm, 90 % improvement of the

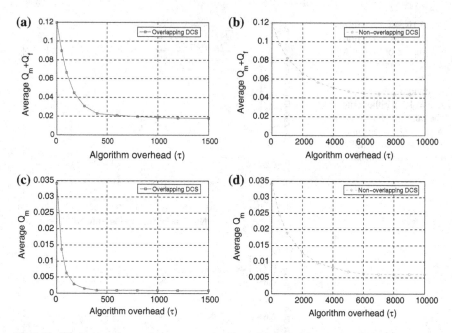

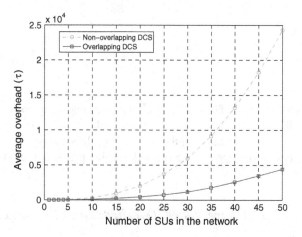

Fig. 3.11 Network sensing performance as a function of the algorithm overhead with network size $N = 30$, power constraint $P_{SU} = 100\,\text{mW}$ and bandwidth constraint $\theta_{SU} = 10$ in both criteria. The false alarm constraint in the Q_m/Q_f criterion is $\alpha = 0.1$

Fig. 3.12 Average overhead as a function of the number of SUs N with power constraint $P_{SU} = 100\,\text{mW}$ and bandwidth constraint $\theta_{SU} = 10$

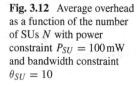

network sensing performance is obtained within the first $300\,\tau$ bits, while for the nonoverlapping algorithm, it takes $4000\,\tau$ bits to achieve the same percentage.

In Fig. 3.12, we show the overhead as a function of the number of SUs N for each algorithm. In this figure, we can see that the overhead of the overlapping algorithm is only about 20 % of the nonoverlapping algorithm, though the overlapping algorithm

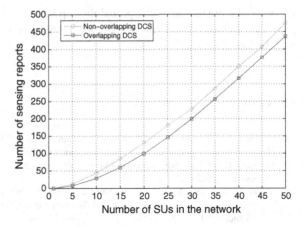

Fig. 3.13 Number of sensing reports as a function of the number of SUs N with power constraint $P_{SU} = 100\,\text{mW}$ and bandwidth constraint $\theta_{SU} = 10$

outperforms the nonoverlapping algorithm in terms of network sensing performance as previously shown. In the nonoverlapping algorithm, the basic operation involves the complete information of two coalitions, and the overhead is produced to check the feasibility of merge operations, as seen in (3.51), even if the operation is not feasible. In the overlapping DCS, the feasibility of switch operation can be locally checked by the SU performing it, and no overhead is produced if the operation is not feasible. Therefore, the overlapping DCS needs less overhead than the nonoverlapping DCS.

In Fig. 3.13, we show the number of sensing reports as a function of the number of SUs N for each algorithm. We see that the overlapping algorithm only needs 90 % sensing reports compared to the nonoverlapping algorithm, which implies a lower system complexity for the cooperative sensing process. In the nonoverlapping approach, an SU joins only one coalition, but it must report to all the members in this coalition. In the overlapping approach, although an SU may join multiple coalitions, it only needs to report to one SU for each coalition it joins, i.e., report to $i \in \mathcal{N}$ for joining coalition $\mathcal{R}_i$. Therefore, the overlapping approach, in which some SUs join multiple coalitions, does not necessarily imply an increase in system complexity.

3.8 Summary

In this chapter, we have discussed a game theoretical approach for distributed cooperative sensing (DCS) with strict power and bandwidth constraints, in which the secondary users can form overlapping coalitions to optimize their spectrum sensing performance. In each coalition, a particular secondary combines the local sensing data from other coalition members and make a cooperative decision. The presented algorithm is proved to converge to a stable outcome within finite iterations. Simulation results show that the overlapping algorithm yields significant performance improvements, decreasing the total error probability up to 25 % in the $Q_m + Q_f$

criterion, the missed detection probability up to 20 % in the Q_m/Q_f criterion, the overhead up to 80 %, and the total report number up to 10 %, compared with the nonoverlapping algorithm.

References

1. E. Hossain, D. Niyato, Z. Han, *Dynamic Spectrum Access and Management in Cognitive Radio Networks* (Cambridge University Press, Cambridge, 2009)
2. B. Wang, K.J.R. Liu, Advances in cognitive radio networks: a survey. IEEE J. Sel. Topics Sig. Process. **5**(1), 5–23 (2011)
3. T. Yucek, H. Arslan, A survey of spectrum sensing algorithms for cognitive radio applications. IEEE Commun. Surv. Tutor. **11**(1), 116–130, first quarter (2009)
4. I.F. Akyildiz, B.F. Lo, R. Balakrishnan, Cooperative spectrum sensing in cognitive radio networks: a survey. Phys. Commun. **4**(1), 40–62 (2011)
5. A. Ghasemi, E.S. Sousa, Collaborative spectrum sensing for opportunistic access in fading environments, in *Proceedings of IEEE International Symposium on New Frontiers in Dynamic Spectrum Access Networks*, Baltimore, MD, Nov. 2005, pp. 131–136
6. C. Sun, W. Zhang, K.B. Letaief, Cooperative spectrum sensing for cognitive radios under bandwidth constraints, in *Proceedings of IEEE Wireless Communications and Networking Conference*, Kowloon, Mar. 2007, pp. 1–5
7. E.C.Y. Peh, Y.C. Liang, Y.L. Guan, Y. Zeng, Optimization of cooperative sensing in cognitive radio networks: a sensing-throughput tradeoff view. IEEE Trans. Veh. Technol. **58**(9), 5294–5299 (2009)
8. W. Zhang, R. Mallik, K. Letaief, Optimization of cooperative spectrum sensing with energy detection in cognitive radio networks. IEEE Trans. Wirel. Commun. **8**(12), 5761–5766 (2009)
9. R. Fan, H. Jiang, A.H. Sayed, Optimal multi-channel cooperative sensing in cognitive radio networks. IEEE Trans. Wirel. Commun. **9**(3), 1128–1138 (2010)
10. Z.M. Fadlullah, H. Nishiyama, N. Kato, M.M. Fouda, Intrusion Detection System (IDS) for combating attacks against cognitive radio networks. IEEE Netw. Mag. **27**(3), 51–56 (2013)
11. G. Ganesan, Y. Li, Cooperative spectrum sensing in cognitive radio, Part I: two user networks. IEEE Trans. Wirel. Commun. **6**(6), 2204–2213 (2007)
12. G. Ganesan, Y. Li, Cooperative spectrum sensing in cognitive radio, Part II: multiuser networks. IEEE Trans. Wirel. Commun. **6**(6), 2214–2222 (2007)
13. W. Saad, Z. Han, T. Basar, M. Debbah, A. Hjorungnes, Coalition formation games for collaborative spectrum sensing. IEEE Trans. Veh. Technol. **60**(1), 276–297 (2011)
14. W. Wang, B. Kasiri, C. Jun, A.S. Alfa, Distributed cooperative multi-channel spectrum sensing based on dynamic coalitional game, in *Proceedings of IEEE Global Telecommunications Conference*, Miami, FL (2010), pp. 1–5
15. O.N. Gharehshiran, A. Attar, V. Krishnamurthy, Dynamic coalition formation for resource allocation in cognitive radio networks, in Proceedings of IEEE International Conference on Communications, Cape Town, South Africa (2010), pp. 1–6
16. R. Yu, Y. Zhang, Y. Liu, S.L. Xie, L.Y. Song, M. Guizani, Secondary users cooperation in cognitive radio networks: balancing sensing accuracy and efficiency. IEEE Wirel. Commun. Mag. **19**(2), 30–37 (2012)
17. Y. Liu, S.L. Xie, R. Yu, Y. Zhang, An efficient MAC protocol with selective grouping and cooperative sensing in cognitive radio networks. IEEE Trans. Veh. Technol. **62**(8), 3928–3941 (2013)
18. H. Li, Cooperative spectrum sensing via belief propagation in spectrum-heterogeneous cognitive radio systems, in *Proceedings of Wireless Communications and Networking Conference*, Sydney, Australia, Apr. 2010, pp. 1–6

19. Z. Zhang, Z. Han, H. Li, D. Yang, C. Pei, Belief propagation based cooperative compressed spectrum sensing in wideband cognitive radio networks. IEEE Trans. Wirel. Commun. **10**(9), 3020–3031 (2011)

20. Q. Wu, D.G.J. Wang, Y. Yao, Spatial-temporal opportunity detection for spectrum-heterogeneous cognitive radio networks: two-dimensional sensing. IEEE Trans. Wirel. Commun. **12**(2), 516–526 (2013)

21. G. Ding, J. Wang, Q. Wu, F. Song, Y. Chen, Spectrum sensing in opportunity-heterogeneous cognitive sensor networks: how to cooperate? IEEE Sens. J. **13**(11), 4247–4255 (2013)

22. A. Malady, C. da Silva, Clustering methods for distributed spectrum sensing in cognitive radio systems, in *Proceedings of IEEE MILCOM* (2008)

23. A.S. Cacciapuoti, I.F. Akyildiz, L. Paura, Correlation-aware user selection for cooperative spectrum sensing in cognitive radio Ad Hoc Networks. IEEE J. Sel. Areas Commun. **30**(2) (2012)

24. A.S. Cacciapuoti, I.F. Akyildiz, L. Paura, Optimal primary-user mobility aware spectrum sensing design for cognitive radio networks. IEEE J. Sel. Commun. **31**(11), 2161–2172 (2013)

25. A.S. Cacciapuoti, M. Caleffi, L. Paura, R. Savoia, Decision maker approaches for cooperative spectrum sensing: participate or not participate in sensing? IEEE Trans. Wirel. Commun. **12**(5), 2445–2457 (2013)

26. W. Saad, Z. Han, M. Debbah, A. Hjorungnes, T. Basar, Coalitional game theory for communication networks. IEEE Signal Process. Mag. Spec. Issue Game Theory **26**(5), 77–97 (2009)

27. Z. Han, D. Niyato, W. Saad, T. Basar, A. Hjorungnes, *Game Theory in Wireless and Communication Networks: Theory, Models and Applications* (Cambridge University Press, Cambridge, 2011)

28. T. Wang, L. Song, Z. Han, W. Saad, Overlapping coalitional games for collaborative sensing in cognitive radio networks, in *Wireless Communications and Networking Conference*, ShangHai, China (2013)

29. T. Wang, L. Song, Z. Han, Z, W. Saad, Distributed cooperative sensing in cognitive radio networks: an overlapping coalition formation approach. IEEE Trans. Commun. **62**(9), 3144–3160 (2014)

30. G. Chalkiadakis, E. Elkind, E. Markakis, N.R. Jennings, Cooperative games with overlapping coalitions. J. Artif. Intell. Res. **39**(1), 179–216 (2010)

31. Z. Zhang, L. Song, Z. Han, W. Saad, Coalitional games with overlapping coalitions for interference management in small cell networks. *IEEE Trans. Wirel. Commun.* **13**(5), 2659–2669 (2014)

32. B. Di, T. Wang, L. Song, Z. Han, Incentive mechanism for collaborative smartphone sensing using overlapping coalition formation games, in *Proceedings of IEEE Global Communications Conference*, Atlanta (2013), pp. 1705–1710

33. Q. Zhao, L. Tong, A. Swami, Y. Chen, Decentralized cognitive MAC for opportunistic spectrum access in Ad Hoc networks: a POMDP framework. IEEE J. Sel. Commun. **25**(3), 589–600 (2007)

34. C. Peng, H. Zheng, B.Y. Zhao, Utilization and fairness in spectrum assignment for opportunistic spectrum access. *ACM Mob. Netw. Appl.* **11**(4), 555–576 (2006)

35. J. Nocedal, S.J. Wright, *Numerical Optimization*, 2nd edn. (Springer, New York, 2006)

36. P. Houze, S.B. Jemaa, P. Cordier, "Common pilot channel for network selection, in *Proceedings of IEEE Vehicular Technology Conference*, Melbourne, Australia (2006), pp. 67–71

37. O. Sallent, J. Perez-Romero, R. Agusti, P. Cordier, Cognitive pilot channel enabling spectrum awareness, in *Proceedings of IEEE International Conference on Communications Workshop*, Dresden, Germany (2009), pp. 1–6

38. Cognitive Wireless Regional Area Network–Functional Requirements, IEEE Std. 802.22 (2006)

Chapter 4
Challenges and Future Works

4.1 Challenges of OCF Games

In the previous chapters, we have shown how to use OCF game approaches to solve complex resource allocation problems in modern cellular networks. Specifically, we consider the interference management problem in heterogenous networks and the cooperative spectrum sensing problem in cognitive radio. However, we point out that there may exists some technical challenges when applying OCF game approaches to other resource allocation problems.

- The cooperative behavior of each coalition member within each coalition must be clarified, e.g., the TDMA coordination for the interference management in heterogeneous networks, so that the coalition value can be well formulated by a specific value function.
- The mutual influence between coalitions must be quantified and restricted, e.g., the inter-coalition interference is restricted between nearby coalitions, so that the computational complexity can be restricted and the coalition-based method can approach the optimal result.
- Practical issues must be considered when applying coalition-based methods, as the corresponding algorithm always require communication between players and coalitions.

4.2 Other Applications

OCF game are quite suitable for modeling the future wireless networks, in which the wireless nodes are dense, self-organizing, and cooperative. In this section, we briefly discuss other potential applications of OCF games and then summarize the applications in Table 4.1.

© The Author(s) 2017
T. Wang et al., *Overlapping Coalition Formation Games
in Wireless Communication Networks*, SpringerBriefs in Electrical
and Computer Engineering, DOI 10.1007/978-3-319-25700-6_4

Table 4.1 Applications of OCF games

Application	Coalition	Coalition Value	Type
Radio resource allocation in Hetnets	RBs from different SBSs	Total throughput of the coalition considering all the interference	K-coalition
Cooperative spectrum sensing in CR	Signaling bits of different SUs to report to an SU	Cooperative sensing performance of the specific SU	K-task
Traffic offloading in multiradio networks	Load traffic of different users distributed to a BS	User performance of the specific BS	K-task
CoMP	The downlink resources of nearby BSs for a cell-edge user	The throughput of the cell-edge user	K-task
Virtual MIMO	Cooperative users forming a virtual MIMO group	The MIMO link throughput	K-coalition
Smartphone sensing	The energy contributed by different smartphones for a task	The task utility	K-task
Subchannel Allocation in NOMA	The power of different subchannels for an end user	The throughput of the end user	K-task
Pilot reuse in Massive MIMO	The users with the same pilot	The total throughput of all users in the coalition	K-coalition

4.2.1 Multiradio Traffic Offloading

Cellular networks are constantly evolving into their next generation. However, the former systems are not entirely replaced by the new systems. In fact, it is expected that different networks will coexist for a long time, and, thus, mobile phones will be multimode terminals that enable communications over different radio access technologies (RATs). In order to fully explore their network investments, the operators must intelligently offload their network traffic over different RATs [1]. Developing such offloading schemes, which must consider the demands and access authorities of different users, the transmitting rates of different technologies, and the deployment and load of different base stations, is quite challenging for a large number of users and base stations. However, one can use the K-task OCF game to model this problem.

In the OCF game model, the mobile users can distribute their traffic into different base stations in different networks. A coalition here represents a base station as well as the traffic devoted from different mobile users. The coalition value can be simply defined as the total throughput of this base station with channel and technology limitations, or a sophisticated function reflecting the user experience, which considers the delay and rate experienced by the users, and the cost and energy efficiency of the network. Using the developed algorithm in Table 1.2, the user traffic can be intelligently distributed among different networks with high network performance in the sense of the defined value function.

4.2.2 Cooperative Communications

In order to increase the performance of cell-edge users, coordinated multipoint (CoMP) transmission has been proposed, in which the signals of multiple base stations are coordinated to serve a cell-edge user [2]. Since there are multiple cell-edge users, the base stations should allocate their radio resources among these users. It is a challenging optimization problem, since the channel conditions, traffic demands, and radio resources are different for different users and base stations. However, we can model this problem using a K-task OCF game. In the OCF game model, the base stations can freely allocate their radio resources to different users, including bandwidth, power, and antenna resources. A coalition represents a cell-edge user as well as the radio resources devoted from different base stations. The coalition value is defined as the throughput of this cell-edge user. Thus, using the developed algorithm in Table 1.2, the radio resources of base stations can be efficiently distributed among different cell-edge users.

Another related application is the cooperation between user devices [3]. In order to increase their transmission rate, nearby users may group together to use virtual MIMO transmissions. The MIMO link rate is generally increasing with the number of cooperated users, while the marginal increase is decreasing due to the increasing distance between different users. Thus, a user may want to allocate its radio resources among different cooperative groups, so as to maximize its individual throughput. This problem can be modeled via a K-coalition OCF game, in which a coalition represents a virtual MIMO group and the coalition value is the MIMO link rate. Using the developed algorithm in Table 1.1, the radio resources of users can be efficiently distributed among different virtual MIMO groups.

4.2.3 Smartphone Sensing

In recent years, smartphones are equipped with more and more sensors. These powerful sensors allow public departments or commercial companies to accomplish large-area sensing tasks via individual smartphones [4]. These tasks often require collecting data in a large area, and thus, a huge number of smartphones may be involved. Based on the task itself and the geographic locations of smartphones, different tasks may require different amount of energy and provide different payoffs for different smartphones. A smartphone user must decide to which tasks he should devote the limited energy. Therefore, we can model this problem with the studied K-task OCF game, in which each coalition represents a task and the energy devoted from different smartphones, and the coalition value is given by the task utility. Using the developed algorithm in Table 1.2, the smartphone users can efficiently allocate their energy into different sensing tasks.

4.2.4 Subchannel Allocation in NOMA

In non-orthogonal multiple access networks, the end users are assigned with over-lapping uplink subchannels by using sparse coding at the user side. Due to the dynamics capacity of different subchannels and different code books, the network performance can be severely influenced by the subchannel allocation policy, which has been proved to be an NP-hard problem [5, 6]. This problem can be formulated as a K-task OCF game, in which the transmission of each user is considered as a task, and each subchannel contributes a part of its available power to multiple tasks. By using the algorithm in Table 1.2, the base station can efficiently allocate its subchannels to different NOMA users.

4.2.5 Pilot Reuse in Massive MIMO

In Massive MIMO systems, the network capacity is limited by the number of available pilot sequences, which is usually insufficient for the future dense networks. Various methods have been proposed to improve the efficiency of pilot sequence reuse via low-intensity BS coordination [7, 8]. This pilot reuse problem can be formulated as a K-coalition OCF game, in which the BSs form cooperative coalitions by devoting the available pilot sequences. Within each coalition, the pilot sequences cannot be reused, and thus, the inner-coalition pilot contamination is eliminated, while the inter-coalition pilot contamination still exists. We can use the algorithm in Table 1.1 to achieve a stable and efficient coalition structure, which indicates how the BSs can coordinate with each other in a large area.

References

1. O. Galinina, S. Andreev, M. Gerasimenko, Y. Koucheryavy, N. Himayat, S.P. Yeh, S. Talwar, Capturing spatial randomness of heterogeneous cellular/WLAN deployments with dynamic traffic. IEEE J. Sel. Areas Commun. **32**(6) (2014)
2. A. Li, R.Q. Hu, Y. Qian, G. Wu, Cooperative communications for wireless networks: techniques and applications in LTE-advanced systems. IEEE Wirel. Commun. **19**(2), 22–29 (2012)
3. S.K. Jayaweera, Virtual MIMO-based cooperative communication for energy-constrained wire-less sensor networks. IEEE Trans. Wirel. Commun. **5**(5), 984–989 (2006)
4. N.D. Lane, Community-aware smartphone sensing systems. IEEE Internet Comput. **16**(3), 60–64 (2012)

5. H. Nikopour, H. Baligh, Sparse code multiple access. in *IEEE 24th International Symposium on Personal Indoor and Mobile Radio Communications (PIMRC)*, London, UK (2013)

6. Y. Saito, Y. Kishiyama, A. Benjebbour, T. Nakamura, A. Li, K. Higuchi, Non-orthogonal multiple access for cellular future radio access, in *IEEE 77th Vehicular Technology Conference (VTC Spring)*, Dresden, Germany (2013)

7. A. Ashikhmin, T.L. Marzetta, Pilot contamination precoding in multi-cell large scale antenna systems, in *Proceedings of the IEEE International Symposium on Information Theory*, MIT Cambridge (2012)

8. H. Yin, D. Gesbert, M. Filippou, Y. Liu, A coordinated approach to channel estimation in large-scale multiple-antenna systems. *IEEE J. Sel. Areas Commun.* **31**(2), 264–273 (2013)

Printed in the United States
By Bookmasters